T0299344

Crafts of Simulation Programming

Crafts of Simulation Programming

E Jack Chen

NEW JERSEY • LONDON • SINGAPORE • BEIJING • SHANGHAI • HONG KONG • TAIPEI • CHENNAI • TOKYO

Published by

World Scientific Publishing Co. Pte. Ltd.
5 Toh Tuck Link, Singapore 596224
USA office: 27 Warren Street, Suite 401-402, Hackensack, NJ 07601
UK office: 57 Shelton Street, Covent Garden, London WC2H 9HE

Library of Congress Cataloging-in-Publication Data
Names: Chen, E. Jack.
Title: Crafts of simulation programming / by E Jack Chen (BASF Corporation, USA).
Description: [Hackensack] New Jersey : World Scientific, 2016. |
Includes bibliographical references and index.
Identifiers: LCCN 2016002837 | ISBN 9789814740173 (hardcover : alk. paper)
Subjects: LCSH: Computer simulation.
Classification: LCC QA76.9.C65 C478 2016 | DDC 303/.3--dc23
LC record available at http://lccn.loc.gov/2016002837

British Library Cataloguing-in-Publication Data
A catalogue record for this book is available from the British Library.

Printed in Singapore

In memory of my grandmother

Preface

Computer simulation is the discipline of studying a wide range of models of real-world systems by numerical evaluation using software designed to imitate the system's operations or characteristics. That is, computer simulation is the process of designing and creating a computerized model of a real or proposed system for the purpose of conducting experiments to give us a better understanding of the behavior of the system under study for a given set of conditions. Simulation output usually consists of one or more random variables because of the stochastic nature of the output data; and output analysis refers to the examination of the data generated by a simulation. Simulation studies have been used to investigate the characteristics of systems, for example, the probability of a machine breakdown.

There are many books on simulation programming with simulation modeling languages, e.g., Arena, ProModel. Those books provide nice introductions of computer simulation and how to build simulation models with high-level languages. On the other hand, we discuss simulation programming in general-purpose languages, namely, C. This approach is highly customizable and flexible, but also painfully tedious and error-prone since models had to be coded pretty much from scratch every time. Nevertheless, by learning to simulate in a general-purpose language, one will have a greater understanding of how simulations actually work. Furthermore, we provide the derivation and basis of various algorithms. That is, we investigate the engine under the hood of the simulation modeling languages. The chapters are organized as follows.

- Chapter 1 discusses the algorithm of generating uniform random numbers, random variates, and some utilities.
- Chapter 2 discusses the effect of sample sizes and stopping rules

for steady-state simulations.

- Chapter 3 presents a procedure to manufacture independent and identically distributed batch means so that classical statistical technique of constructing confidence intervals can be used.
- Chapter 4 reviews the fundamentals of order statistics and how they are applied in simulation.
- Chapter 5 discusses order statistics from correlated normal random variables and its relationship with ranking and selection.
- Chapter 6 presents procedures to estimate multiple quantiles via empirical histograms of the underlying distributions.
- Chapter 7 presents procedures to construct metamodels, which can be used as a surrogate to study the underlying system.
- Chapter 8 presents procedures to estimate density of underlying distributions.
- Chapter 9 presents procedures to compare two normal populations.
- Chapter 10 reviews ranking and selection procedures as well as multiple comparisons.
- Chapter 11 compares the indifference-zone selection procedures with optimal computation allocation strategy.
- Chapter 12 presents some insight of using common random numbers with selection procedures to increase the probability of correct selection.
- Chapter 13 discuss using parallel and distributed simulation to increase the capacity of simulation studies.
- Chapter 14 reviews multiple-objective selection procedures and a Pareto set.
- Chapter 15 reviews a generic selection-with-constraints procedure. The constraints can be based on a standard or a control.

E. J. Chen

Contents

Chapter 1

Basic Simulation Programming

1.1 Introduction

Computer simulation is the process of designing and creating a computerized model of a real or proposed system for the purpose of conducting experiments to give us a better understanding of the behavior of the system under study for a given set of conditions. Simulation studies have been used to investigate the characteristics of systems, to assess and analyze risks, for example, the probability of a machine breakdown.

Simulation studies are typically proceed by transforming in a more or less complicated way of a sequence of numbers between 0 and 1 produced by a pseudorandom generator into an observation of the measure of interest. A facility for generating sequences of pseudorandom numbers is a fundamental part of computer simulation systems.

1.2 Random Numbers Generators

A portable set of software utilities is described for uniform random-number generation. It provides for multiple generators (streams) running simultaneously, and each generator (stream) has its sequence of numbers partitioned into many long disjoint contiguous substreams. Simple procedure calls allow users to make any generator "jump" ahead/back v steps (random numbers). Implementation issues are discussed. An efficient and portable code is also provided to implement the package.

A collection of random variables $x_1, x_2, \ldots, x_n$ is a random sample if they are independent and identically distributed (i.i.d.). True random numbers cannot be produced by a deterministic algorithm, and hence, random numbers generated by using a recursive equation are referred to as *pseudo-*

random numbers. Usually, in practice, such deterministic algorithms produce a deterministic sequence of values, but externally these values should appear to be drawn independently from a uniform distribution between 0 and 1 (i.e., $U(0,1)$). Furthermore, multiple independent streams of random numbers are often required in simulation studies, for instance, to facilitate synchronization for variance-reduction purposes, and for making independent replications. A random number generator (RNG) is an algorithm that starting from an initial seed (or seeds), produces a stream of numbers that behaves as if it was a random sample when analyzed using statistical tests. The RNG is closely related to the Deterministic Random Bit Generators (DRBGs). See [L'Ecuyer (1990)] and references therein for more information on RNGs. We describe a portable set of software utilities for uniform random-number generation. It provides for multiple generators (streams) running simultaneously, and each generator (stream) has its sequence of numbers partitioned into many long disjoint contiguous substreams. The basic underlying generator CMRG (Combined Multiple Recursive Generator) combines two multiple recursive random number generators with a period length of approximately 2^{191} ($\approx 3.1 \times 10^{57}$), good speed, and excellent theoretical properties. See [L'Ecuyer et al. (2002)].

There are a number of methods for generating the random numbers, of which the most popular are the congruential methods (mixed, multiplicative, and additive). The (mixed) linear congruential generators (LCGs) are defined by

$$x_i = (ax_{i-1} + c) \bmod m, u_i = x_i/m, x_0 \in \{1, \cdots, m-1\}, i > 0.$$

Here m (the modulus) is a positive integer (usually a very large primary number), a (the multiplier) $\in \{0, 1, \cdots, m-1\}$ and c (the increment) is a nonnegative integer. This mathematical notation signifies that x_i is the remainder of $(ax_{i-1} + c)$ divided by m. Hence, $x_i \in \{0, 1, \cdots, m-1\}$. Thus, random variable u_i is a uniform 0, 1 variable. Note that $x_{i+v} = (a^v x_i + c(a^v - 1)/(a-1)) \bmod m$. Hence, every x_i is completely determined by m, a, c, and x_0. The sequence x_i repeats once it returns to a previously visited value. The *period* of a generator is the length of a generated steam before it begins to repeat. If $u_0 = u_p$ (where $p > 0$), then the length p is called the period . The longest possible period for a LCG is m, i.e., m represents the desired number of different values that could be generated for the random numbers. Hence, the modulus m is often taken as a large prime number close to the largest integer directly representable on the computer (i.e., equal or near $2^{31} - 1$ for 32-bit computers). If $p = m$, we say that the generator has full period. The required conditions on how to choose m, a, and c so that the corresponding LCGs will have full period are known, see [Knuth (1998)] or [Law (2014)].

When $c > 0$, the LCGs are called mixed LCGs. When $c = 0$, $x_i = ax_{i-1} \bmod m$, $u_i = x_i/m, x_0 \in \{1, \cdots, m-1\}, i > 0$. These LCGs are called multiplicative LCGs. Note that if $x_i = 0$, then all subsequent x_i are identically 0. Thus, the longest possible period for a multiplicative LCG is $m - 1$. Furthermore, $x_{i+v} = a^v x_i \bmod m$. Most experts now recognize that small LCGs with moduli around 2^{31} or so should no longer be used as general-purpose random-number generators. Not only can one exhaust the period in a few minutes on a PC (personal computer), but more importantly the poor structure of the points can dramatically bias simulation results for sample sizes much smaller than the period length.

One way of extending the basic LCG is to combine two or more LCGs through summation. Another way of extending the basic LCG is to use a higher-order recursion. A multiple recursive random number generator (MRG) , which goes from integer to integer according to the recursion

$$x_i = (a_1 x_{i-1} + \cdots + a_k x_{i-k}) \bmod m, u_i = x_i/m.$$

A seed $x_0, \ldots, x_{k-2}, x_{k-1} \in \{1, \ldots, m-1\}$, where i, k, and m are positive integers, and $a_1, \ldots, a_k \in \{0, 1, \ldots, m-1\}$. To increase the efficiency and ease the implementation, the MRG algorithm usually set all but two a_i's to 0. Furthermore, the nonzero a_i should be small. However, these conditions are generally in conflict with those required for having a good lattice structure and statistical robustness. See [Law (2014)] on the lattice structure of pseudorandom numbers. The longest possible period for a MRG is $m^k - 1$ [L'Ecuyer (1996)]. Other way of extending the basic LCG is the additive congruential RNG (ACRON). The ACRON sets $a = 1$ and replace c by some random number preceding x_i in the sequence, for example, x_{i-1} (so that more than one seed is required to start calculating sequence). [Wilkrmaratna (2008)] indicated that the ACRON is a special case of a multiple recursive generator.

Other classes of RNGs are available, e.g., twisted generalized feedback shift register generators [Matsumoto and Nishimura (1998)]. The RNGs discussed so far are by computational methods. Another approach is by physical methods, e.g., [Kanter et al. (2010)]. The output based on a physical method is usually unpredictable. Thus, this class of random bit generators is commonly known as non-deterministic random bit generators (NRBGs). Those physical devices are unnecessary and unpractical for

Monte Carlo methods, because deterministic algorithmic methods are much more convenient. See [Law (2014)] for more information on other methods of generating random numbers.

We discuss the algorithm and implementation of [L'Ecuyer (1999)] combined multiple recursive random number generator. A combination of generators can have a much longer period than any of its components. Furthermore, with well-chosen parameters, the CMRG has good structural properties and passes many statistical tests.

1.2.1 *Basic Generators*

A newly employed generator is the combined multiple recursive random number generator, which goes from integer to integer according to the recursion

$$x_{j,n} = (a_{j,1}x_{j,n-1} + \cdots + a_{j,k}x_{j,n-k}) \bmod m_j; \quad \text{for} \quad j = 1, \ldots, J \tag{1.1}$$

$$z_n = \sum_{j=1}^{J} (\delta_j x_{j,n}) \bmod m_1 \tag{1.2}$$

$$u_n = \begin{cases} z_n/(m_1+1) & \text{if } z_n > 0 \\ m_1/(m_1+1) & \text{if } z_n = 0. \end{cases} \tag{1.3}$$

Here J is the number of MRGs used in the CMRG, j, n, k, and m_j are positive integers, and each $a_{j,k}$ belongs to $Z_m = \{0, 1, \ldots, m - 1\}$, δ_j are integers, and the greatest common divisor of δ_j and m_j is one for each j. The transformation of u_n makes sure that u_n is never equal to 0 or 1 (otherwise, trouble may arise, e.g., when taking the logarithm to generate an exponential random variate). Assume that the m_j are distinct primes and that each recurrence j has a primitive characteristic polynomial, and thus period length $\rho_j = m_j^k - 1$ and the longest possible period for a CMRG is $\rho_1 \cdots \rho_J / 2^{J-1}$. The combined generator is a close approximation to a single MRG with a modulus equal to the product of the moduli of the components MRGs. Thus, the CMRG has the advantages associated with a larger modulus while permitting an implementation using smaller values. The CMRG combining multiple recursive sequences provides an efficient way of implementing random-number generators with long periods and good structural properties. If the parameters are well chosen, such generators are statistically far more robust than simple linear congruential generators that fit into a single computer word [L'Ecuyer (1999)].

1.2.2 *The Need for Multiple Substreams*

Many disjoint random number subsequences are often required in simulation studies, for instance, 1) to make independent replications and/or; 2) to associate distinct "streams" of random numbers with different sources of randomness in the system to facilitate synchronization for variance reduction. To produce such "streams," different seeds (values of vector S_i) must be obtained far enough apart in the sequence to insure that the streams do not overlap. Selection of seeds should also consider statistical properties between streams, such as apparent independence. In other words, given any X_i (with seed S_i) and positive integer v, there should be a quick way to compute X_{i+v} (with seed S_{i+v}) without generating all intermediate values. The availability of efficient jump-ahead methods is very useful because it permits one to partition the RNG sequence into long disjoint stream and substreams of random numbers. Most packages offer no facility for jumping ahead directly from X_i to X_{i+v} or to compute distant seeds efficiently. Many simulation languages offer a limited number of streams, all based on the same generator, but using fixed starting seeds set say 100,000 values apart. This provides relatively low flexibility. Suppose, for instance, that you want to perform independent pairs of replications with common random numbers across the configurations (i.e., between any two runs of the same pair) in order to compare two different configurations of a system. To insure proper synchronization, you want every generator to start from the same seed in both runs of the same pair. However, in general, these two runs will make a different number of calls to a generator, and programming "tricks" should be used to skip a proper amount of random numbers to resynchronize the generators for the next pair without overlap in the random number streams. This requires extra programming effort and could be error prone. Good software tools should ease the programmer's task in that respect. A simple procedure call should permit resetting a generator to previous seed or jumping ahead to a new seed for the next run. Of course, the sequence of "new seeds" (one per run) should be the same of both configurations for the system. Implementing such tools requires efficient "jumping ahead" facilities, which in turn ask for efficient procedures to compute $(a \times s) \bmod m$, where a and s are positive integers.

1.2.3 Computing $(a \times s) \bmod m$

Consider a 32-bit computer on which all integers between $-2^{32}-1$ and $2^{32}-1$ (exclusive) are well represented. We want to compute $(a \times s) \bmod m$, where a, s, and m are positive integers smaller than $2^{32}-1$. Without loss of generality, we assume that $a < m$ and $s < m$ (if not, replace a and s by $a \bmod m$ and $s \bmod m$, respectively). In order to keep seeds with 32-bit precision, all operations can not produce any number greater than 2^{53} (the IEEE-754 standard, i.e., the floating-point numbers have at least 53 bits of precision for the mantissa). Therefore, special algorithm is needed to compute $(a \times s) \bmod m$, where a and s are less than 2^{32} (around 4.3×10^9). The following algorithm is used. Let $a = a_1 \times 2^{17} + a_2$ so $a \times s = a_1 \times s \times 2^{17} + a_2 \times s$. Therefore,

$$(a \times s) \bmod m = (((a_1 \times s) \bmod m) \times 2^{17} + a_2 \times s) \bmod m,$$

where $a_1 < 2^{15}$, so $a_1 \times s < 2^{47} (= 2^{15} \times 2^{32})$, and $a_2 < 2^{17}$, so $a_2 \times s < 2^{49}$ $(= 2^{17} \times 2^{32})$. Because $z = (a_1 \times s) \bmod m < 2^{32}$, we have $z \times 2^{17} < 2^{49}$ so that all the intermediate terms in the above computations are less than 2^{53}. Therefore, all seed values will have exact accuracy.

1.2.4 Computing the Jumping Matrices

The initial state of substreams can be computed easily if jumping-ahead facilities are available for the individual MRG components; that is, if an efficient algorithm is available for computing the state of the MRG v steps ahead of the current one, for large values of v. [L'Ecuyer (1990)] explained one way of doing that, based on the fact that the MRG can be viewed as a LCG in matrix form, whose state is a k-dimensional vector and whose multiplier is a $k \times k$ matrix A. To jump ahead by v values, just multiply the current state by $(A^v \bmod m)$. The matrix by $(A^v \bmod m)$ can be pre-computed in time $O(\log v)$, using the divide-and-conquer algorithm [Knuth (1998)]. The divide-and-conquer algorithm uses the following recursion:

$$A^v \bmod m = \begin{cases} A & \text{if } v = 1; \\ A(A^{v-1} \bmod m) \bmod m & \text{if } v > 1,\ v \text{ odd}; \\ (A^{v/2} \bmod m)(A^{v/2} \bmod m) \bmod m & \text{if } v > 1,\ v \text{ even}. \end{cases}$$

That is, for the MRG random number generator, X_{i+v} can be computed directly from X_i using

$$X_{i+v} = (A^v X_i) \bmod m = (A^v \bmod m) X_i \bmod m \tag{1.4}$$

where

$$A = \begin{pmatrix} 0 & 1 & \cdots & 0 \\ \vdots & \vdots & \ddots & \vdots \\ 0 & 0 & \cdots & 1 \\ a_k & a_{k-1} & \cdots & a_1 \end{pmatrix} \tag{1.5}$$

is an invertible $k \times k$ matrix and

$$X_i = \begin{pmatrix} x_i \\ x_{i+1} \\ \vdots \\ x_{i+k-1} \end{pmatrix} \tag{1.6}$$

is a $k \times 1$ vector.

When A has this special structure, the first $k - 1$ components of X_i are obtained by shifting the last $k - 1$ components of X_{i-1}, and the last component of X_i is a linear combination of the components of X_{i-1} according to the MRG recursion [L'Ecuyer (1990)]. [Kroese et al. (2011)] point out that an MRG can be interpreted and implemented as a matrix multiplicative congruential generator.

1.2.5 *A Random Number Package*

We now propose a portable set of utilities for random number generation. We consider a basic underlying generator of period p. Let s_0 be the basic seed (initial state) for this generator and $s_1, s_2, \ldots$ be its sequence of successive states. Let T denote the transition function of the generator, that is, the operator $T : S \to S$ such that $T(s_i) = s_{i+1}$, and T^q its q-fold composition ($T^q(s_i) = s_{i+q}$). Starting from states $I_1 = s_0, I_2 = s_q = T^q(I_1), I_3 = s_{2q} = T^q(I_2), \ldots, I_G = s_{(G-1)q} = T^q(I_{G-1})$, respectively. Each of these G sequences corresponds to a "stream", with length q. Note that q needs to be a very large number, say no smaller than 10^9. Hence, there will be G virtual RNGs. I_g is called the initial seed of stream g. At any moment during program execution, stream g is in some state C_g, say, in its subsequence number r, that is, such that $C_g = T^{r-1}(I_g)$. We call C_g the current state of stream g.

If the product of $a_{j,i}(m_j - 1)$ is less than 2^{53}, then the integer $a_{j,i}$ $x_{j,i}$ is always represented exactly in floating point on a 32-bit computer that supports the IEEE floating-point arithmetic standard, with at least 53 bits of precision for the mantissa. The generator can then be implemented directly in floating-point arithmetic, which is typically faster than an integer arithmetic implementation. On the other hand, with this implementation, the state of the generator is represented over $64k$ J bits, as opposed to $32k$ J bits when the $x_{j,i}$ are represented as 32-bit integers.

The mrg32k3a of [L'Ecuyer (1999)] implemented a CMRG with 2 components ordered 3, whose coefficients satisfy above condition. The moduli and coefficients are $m_1 = 2^{32} - 209$, $a_{1,1} = 0$, $a_{1,2} = 1403580$, $a_{1,3} = -810728$, $m_2 = 2^{32} - 22853$, $a_{2,1} = 527612$, $a_{2,2} = 0$, $a_{2,3} = -1370589$. Its period length is $\rho = (m_1^3 - 1)(m_2^3 - 1)/2 \approx 2^{191} \approx 3.1 \times 10^{57}$. This implementation set $\delta_1 = -\delta_2 = 1$. The parameters have been chosen so that the period is long, a fast implementation is available, and the generator performs well with respect to the spectral test. Before this procedure is called for the first time, one must initialize the global variables $s10, s11, s12$ to (exact) non-negative integer values less than m_1 and not all zero, and s20, s21, s22 to non-negative integers less than m_2 and not all zero. The vectors (s10, s11, s12) and (s20, s21, s22) are the initial values of $(x_{1,0}, x_{1,1}, x_{1,2})$ and $(x_{2,0}, x_{2,1}, x_{2,2})$, respectively. They constitute the seed. Therefore, in this implementation

$$x_{1,n} = (1403580.0x_{1,n-2} - 810728.0x_{1,n-3}) \bmod 4294967087$$

$$x_{2,n} = (527612.0x_{2,n-1} - 1370589.0x_{2,n-3}) \bmod 4294944443$$

$$Z_n = (x_{1,n} - x_{2,n}) \bmod 4294967087$$

$$u_n = \begin{cases} z_n/4294967088 & \text{if } z_n > 0 \\ 4294967087/4294967088 & \text{if } z_n = 0. \end{cases} \tag{1.7}$$

The matrix A_1 for $(x_{1,0}, x_{1,1}, x_{1,2})$ to jump ahead one step is

$$A_1 = \begin{pmatrix} 0 & 1 & 0 \\ 0 & 0 & 1 \\ -810728 & 1403580 & 0 \end{pmatrix}$$

and the matrix A_2 for $(x_{2,0}, x_{2,1}, x_{2,2})$ to jump ahead one step is

$$A_2 = \begin{pmatrix} 0 & 1 & 0 \\ 0 & 0 & 1 \\ -1370589 & 0 & 527612 \end{pmatrix}.$$

The following is an implementation of the CMRG (L'Ecuyer's mrg32k3a) in C.

```
#define norm 2.328306549295728e-10        // 1.0/(m1+1)
#define norm2 2.328318825240738e-10       // 1.0/(m2+1)
#define m1 4294967087.0
#define m2 4294944443.0

// the initial seed
double s[2][3] = {
{0.0,0.0,1.0},
{0.0,0.0,1.0}
};

double MRG32k3a()
{
long k;
double p;

p = 1403580.0 * s[0][1] - 810728.0 * s[0][0];
k = long ( p / m1); p -= k*m1; if (p < 0.0) p += m1;
s[0][0] = s[0][1]; s[0][1] = s[0][2]; s[0][2] = p;

p = 527612.0 * s[1][2] - 1370589.0 * s[1][0];
k = long ( p / m2); p -= k*m2; if (p < 0.0) p += m2;
s[1][0] = s[1][1]; s[1][1] = s[1][2]; s[1][2] = p;

if (s[0][2] <= s[1][2])
    return ((s[0][2] - s[1][2] + m1) * norm);
else
    return ((s[0][2] - s[1][2]) * norm);
}
```

1.2.6 *Jumping Backward*

The jump-back matrix also exists. This is true because each component has a primitive characteristic polynomial (a necessary condition). Given a vector S_n, one may jump back v steps to S_{n-v}. For example, the jump-back-one-step matrix B_j is $A_j^{\rho_j - 1} \bmod m_j$, where ρ_j is the period of the jth MRG of the CMRG. As one might expect, $A_j^v \times B_j^v \bmod m_j$ turns out to be the identity matrix I, for $v = 1, 2, \ldots, \rho_j$. This equation is consistent with the intuition that $A_j^{\rho_j - 1} \bmod m_j = I$. The matrix B_j generates the same stream but in reverse order. As we can see from the new recursion below, the values of the parameters $b_{i,j}$ are much larger than the original ones, where $b_{i,j}$ are the parameters of the new CMRG. Furthermore, we no longer have $b_{i,j}(m_j - 1) < 2^{53}$ for all b_{ij}. Therefore, the original recursion provides a more efficient generator. The matrix B_1 for S_{1n} to jump back one step (i.e., jump ahead $\rho_1 - 1$ step) is

$$B_1 = \begin{pmatrix} 184888585 & 0 & 1945170933 \\ 1 & 0 & 0 \\ 0 & 1 & 0 \end{pmatrix}$$

and the matrix B_2 for S_{2n} to jump back one step (i.e., jump ahead $\rho_2 - 1$ step) is

$$B_2 = \begin{pmatrix} 0 & 360363334 & 4225571728 \\ 1 & 0 & 0 \\ 0 & 1 & 0 \end{pmatrix}.$$

The reverse stream follows the recursion

$$\begin{aligned} x_{1,n} &= (184888585 \times x_{1,n+1} + 1945170933 \times x_{1,n+3}) \bmod 4294967087, \\ x_{2,n} &= (360366334 \times x_{2,n+2} + 4225571728 \times x_{2,n+3}) \bmod 4294944443 \\ &= (360366334 \times x_{2,n+2} - 69372715 \times x_{2,n+3}) \bmod 4294944443. \end{aligned}$$

1.3 Examples of Using Random Number Generator

In the following example, we illustrate how to use the functions in the mrg32k32 RNG software package. The initial seed of the main generator s_0 is the starting point of the first stream (i.e., the first state) I_1. In the proposed package, the initial seed is set to the default value (12345, 12345, 12345, 12345, 12345, 12345), but this value can be changed by the user (via `SetPackageSeed`). Each time a new `RngStream` object is created, its starting point (initial seed) I_g is set $v = 2^{127}$ steps ahead of the starting point of the last created object. A vector named `nextSeed` is used to keep the seed values of the next created `RngStream` object (stream). For example, the declaration "`RngStream g;`" creates a stream with I_g equal to `nextSeed` and advances `nextSeed` by v steps. Because the initial seed for each `RngStream` object is computed dynamically, no pre-computed list of seeds is needed.

The methods `Reset*` reset a given stream either to its initial state, or to the beginning of its current substream, or to the beginning of its next substream. The method `GetState` returns the state of a stream. One can change the seed of a given stream, without modifying that of other streams, by invoking `SetSeed` or `AdvanceState`. However, after calling `SetSeed` for a given stream, the initial states of the different streams are no longer spaced v values apart. Therefore, this method should be used *only in exceptional cases.* The methods `Reset*` suffices for almost all applications. The method `RandU01` generates the uniform (pseudo)random numbers with the seed as an input parameter and advance the state of the seed by one step.

Example 1 shows how we generate a list of ten seed vectors. We set the package seed to {327612383, 317095578, 14704821, 884064067, 1017894425, 16401881} by calling `SetPackageSeed` before instantiating any `RngStream` object. Note that if `SetPackageSeed` is not executed, {12345, 12345, 12345, 12345, 12345, 12345} will be used as the default package seed. The declaration "`RngStream RngObj`" is inside the `for` loop on i. Therefore, the instance of the object dies before the next iteration, i.e., the object `RngObj` is not available outside the `for` loop on i. Each declaration will create the `RngObj` instance with $I_g = C_g =$ `nextSeed` and advance `nextSeed` by 2^{127} steps. Thus, the output seeds will be 2^{127} apart. The procedure "WriteState()" prints the values of the seed.

```
#include "RngStream.h"

int main ()
{
   unsigned long seed[6] =
    { 327612383, 317095578, 14704821,
      884064067, 1017894425, 16401881 };
   RngStream::SetPackageSeed (seed);

   for (int i = 1; i <= 10; ++i) {
      RngStream RngObj;
      RngObj.WriteState();
   };
}
```

Example 2 shows how to apply some of the utilities supplied in the package. The declarations "`RngStream RngObj1`" and "`RngStream RngObj2`" will create the RNG objects with

$$I_g = C_g = \{12345, 12345, 12345, 12345, 12345, 12345\}$$

and

$$I_g = C_g =$$

$$\{3692455944, 1366884236, 2968912127, 335948734, 4161675175, 475798818\},$$

respectively, i.e., 2^{127} steps apart. The first RNG object `RngObj1` may be dedicated to generate inter-arrival times while the second RNG object `RngObj2` may be dedicated to generate service times, for some queueing system to be simulated. We generate five inter-arrival times and five service times, then move each RNG to its next substream. This is repeated ten times, thus yielding ten vectors, each containing five inter-arrival times and five service times. Moreover, these ten vectors will be exactly the same at iterations $r = 0$ and $r = 1$ of the outer `for` loop, because the statements "`RngObj1.ResetStartStream`" and "`RngObj2.ResetStartStream`" will reset the current seeds C_g of both RNG objects to the initial seed I_g.

The function "double Rand()" returns a (pseudo)random number from the uniform distribution over the interval $(0, 1)$, after advancing the state of the internal seed by one step.

```
#include <math.h>
#include "RngStream.h"

int main ()
{
   RngStream RngObj1;
   RngStream RngObj2;
   double interArrival;
   double serviceTime;

   for (int k = 0; k <= 1; ++k) {
      for (int j = 1; j <= 10; ++j) {
         for (int i = 1; i <= 5; ++i) {
            interArrival = -log(1.0 - RngObj1.Rand());
            serviceTime = -0.9 * log(1.0 - RngObj2.Rand());
         };
         RngObj1.ResetNextSubstream ();
         RngObj2.ResetNextSubstream ();
      };
      RngObj1.ResetStartStream ();
      RngObj2.ResetStartStream ();
   };
}
```

1.4 Nonuniform Random Variates

To simulate the underlying system under study, various nonuniform random variates are required. These random variates are used to simulate the observations from the desired distribution. Random variates from distributions other than the uniform over [0,1] are generated by applying further transformations to the output values u_i of the uniform RNG. There is a vast collection of literature on how to generate random variates from a given distribution in order to run the simulation model. One of the popular references is [Law (2014)], where many further references on this subject is listed. The simplest way of generating a random variate X with distribution function F is using the inverse transformation. Based on the fact that $y = F(x) \sim U(0,1)$, where '$\sim$' denotes 'is distributed as', we generate a random variate $y \sim U(0,1)$ and return the value $x = F^{-1}(y)$ as a variate with distribution F. Here F^{-1} is the inverse distribution function (or quantile function) and is defined by

$$F^{-1}(y) = \inf\{x : F(x) \geq y\}.$$

The cumulative distribution function (cdf) of the inverse transform $F^{-1}(y)$ is given by

$$\Pr(F^{-1}(y) \leq x) = \Pr(y \leq F(x)) = F(x).$$

Figure 1.1 illustrates the algorithm. For the discrete case the inverse-

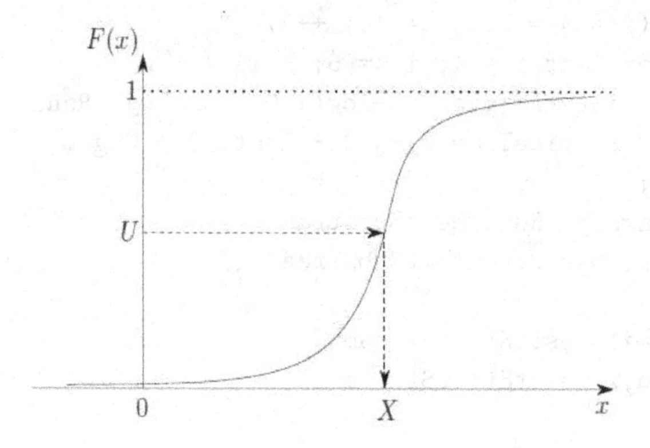

Fig. 1.1 Inverse-transform method for continuous variables

transform method can be done by finding the smallest positive integer k such that $F(x_k) \geq y$ and return $X = x_k$. Here the discrete random variable X taking values $x_1 < x_2 < \cdots$ with probabilities $p_1, p_2, \cdots$ and $\sum p_i = 1$. For more information of using inverse transform to generate random variates, see [Law (2014)]. Other methods are available when F^{-1} is difficult or expensive to compute, e.g., composition, convolution, acceptance-rejection, and special properties, see [Kroese et al. (2011)].

1.4.1 *Random Variates of Various Distributions*

In this section, we list the source code of generating various nonuniform random variates.

```
/*-------------------------------------------------------------*/
/* Remark: Implementation of Marsaglia and Bray (1964)'s      */
/*         polar method of generating N(0,1) random           */
/*         variates.                                          */
/*                                                            */
double RandStdNormal(double seed[6])
{
double v1, v2, temp;
double w = 2.0;
static double even = 0;

    if ( even ) {
      temp = even;
      even = 0;
      return (temp);
    }

    while ( w > 1 ) {
      v1 = (2 * RandU01(seed)) - 1;
      v2 = (2 * RandU01(seed)) - 1;
      w = pow(v1, 2) + pow(v2, 2);
    }

    temp = sqrt (-2 * log (w) / w);
    even = v2 * temp;

    return (v1 * temp);
}

double RandNormal(double seed[6], double mean, double variance)
{
    return (RandStdNormal(seed)*sqrt(variance) + mean);
}



/*-------------------------------------------------------------*/
/* Remark: Implementation of generating exponential           */
/*         distribution with parameter beta, i.e.,            */
```

```
/*          expon(beta).                                      */
/*                                                            */
double RandExpon(double seed[6], double beta)
{
     return (-beta * log(RandU01(seed)));
}


/*-----------------------------------------------------------*/
/* Remark: Implementation of generating erlang distribution */
/*         with parameters beta, alpha.                     */
/*                                                           */
double RandErlang(double seed[6], double alpha, double beta)
{
double u=0.0;
int j;

    for (j = 1; j <= floor(beta); ++j)
        u += expon(seed, alpha);

    return (u);
}


/*-----------------------------------------------------------*/
/* Remark:  Implementation of Cheng (1977)'a algorithm of   */
/*          generating gamma distribution                   */
/*          with parameters alpha, beta.                    */
/*                                                           */
#define M_E  2.718281828
double RandGamma(double seed[6], double alpha, double beta)
{
double a = 1 / sqrt(2*alpha-1);
double b = alpha - log(4.0);
double q = alpha + 1/a;
double theta = 4.5;
double d = 1+log(4.5);
```

```
double u1 = RandU01(seed);
double u2 = RandU01(seed);

double v = a * log(u1/(1-u1));
double y = alpha * pow(M_E,v);
double z = pow(u1,2)*u2;
double w = b + q*v - y;

    if (w + d >= theta * z)
        return (y*beta);

    if (w >= log(z))
        return (y*beta);

    return (RandGamma(seed, alpha, beta));
}


/*----------------------------------------------------------*/
/* Remark:  Implementation of generating beta distribution  */
/*          with parameters alpha, beta.                     */
/*                                                           */
double RandBeta(double seed[6], double alpha, double beta)
{
double y1 = RandGamma(seed, alpha, 1);
AdvanceState(seed,127,0)
double y2 = RandGamma(seed, beta, 1);

    return (y1/(y1+y2));
}


/*----------------------------------------------------------*/
/* Remark: Implementation of generating chi square          */
/*         distribution with parameter df.                   */
/*                                                           */
double RandChi(double seed[6], int df)
{
```

```
    return (RandGamma(seed, (1.0*df)/2, 2));
}


/*-----------------------------------------------------------------*/
/* Remark: Implementation of generating weibull                   */
/*         distribution with parameters alpha, beta.              */
/*                                                                */
double RandWeibull(double seed[6], double alpha, double beta)
{
    return (alpha * pow(-1*log(RandU01(seed)), 1.0/beta));
}


/*-----------------------------------------------------------------*/
/* Remark:  Implementation of generating t distribution           */
/*          with parameter df.                                    */
/*                                                                */
double RandT(double seed[6], int df)
{
double y = RandNormal(seed, 0, 1);
AdvanceState(seed,127,0)
double z = RandChi(seed, df);

    return (y/sqrt(z/df));
}
```

1.4.2 *Correlated Random Variates*

In this section, we list the source code of generating various correlated random variates.

```
/*-----------------------------------------------------------------*/
/* Remark: Implementation of generating steady-state              */
/*         autoregressive AR1 distribution with mean mu and */
/*         correlation coefficient phi.  The error term has */
/*         a N(0,1) distribution.                                 */
/*   parameters: seed is the seed for the uniform number          */
```

```
/*                generator                                  */
/*   phi is the correlation coefficient                      */
/*   mu is the expected value                                */
/*   x0 is the current value                                 */
/*                                                           */
double ar1(double seed[6], double phi, double mu, double* x0)
{
     *x0 = mu + phi*(*x0 - mu) + RandNormal(seed, 0.0, 1.0);

     return (*x0);
}


/*-----------------------------------------------------------*/
/* Remark: Implementation of generating steady-state Moving */
/*         Average MA1 distribution with correlation         */
/*         coefficient phi, mean mu previous noise e, where */
/*         e has a N(0,1) distribution.                      */
/*   parameters: seed is the seed for the uniform number    */
/*                generator                                  */
/*   phi is the correlation coefficient                      */
/*   mu is the expected value                                */
/*   e is a normal(0,1) white noise                          */
/*                                                           */
double ma1(double seed[6], double phi, double mu, double *e)
{
double temp = RandNormal(seed, 0.0, 1.0);
double x0 = temp + mu + phi * (*e);

     *e = temp;

     return (x0);
}


/*-----------------------------------------------------------*/
/* Remark: Implementation of generating the waiting time of */
/*        M/M/1 delay in queue with the arrival rate         */
```

```
/*          lambda and server rate mu.                          */
/*    parameters: seed is the seed for the uniform number      */
/*                generator                                     */
/*    mu is service rate                                        */
/*    lambda is arrival rate                                    */
/*    w is the waiting time of the previous customer            */
/*                                                              */
double mm1que(double seed[6], double mu,
              double lambda,double* w)
{
double a = RandExpon(seed,1.0/lambda);
AdvanceState(seed,127,0);
double s = RandExpon(seed,1.0/mu);

     *w += s - a;
     if (*w  < 0) *x0  = 0.0;

     return (*w);
}
```

1.5 Utilities

In this section, we list several utilities that are often needed for simulation studies. [Hastings (1955)] lists approximations of commonly used distributions. We list the implementations of computing cdf of standard normal distribution, quantile of normal, t, and chi-square distributions as well as procedures to compute variance.

1.5.1 *Numerical Approximation of Normal Distribution*

There is no explicit closed form solution of normal cumulative distribution function. The following is a numerical approximation proposed by [Abramowiz and Stegun (1964)].

```
/*------------------------------------------------------------*/
/* Remark: Compute the value of standard normal probability */
/*          density function                                  */
/*    parameters: x is the location                           */
```

```
/*                                                              */
#define M_E  2.718281828
#define M_PI 3.14159265358979
double pdf_normal(double x)
{
   return ( pow(M_E, -0.5*pow(x,2)) / sqrt(2 * M_PI) );
}



/*--------------------------------------------------------------*/
/* Remark: Compute the value of standard normal cdf             */
/*   parameters: z is the quantile                              */
/*                                                              */
double cdf_normal(double z)
{
   if (z > 6.0) return (1.0);  // to guard against overflow
   if (z < -6.0) return (0.0);

   double b1 = 0.31938153;
   double b2 = -0.356563782;
   double b3 = 1.781477937;
   double b4 = -1.821255978;
   double b5 = 1.330274429;
   double p = 0.2316419;
   double c2 = 0.3989423;

   double a=fabs(z);
   double t = 1.0/(1.0+a*p);
   double b = c2*exp((-z)*(z/2.0));
   double n = ((((b5*t+b4)*t+b3)*t+b2)*t+b1)*t;
   n = 1.0-b*n;
   if (z < 0.0) n = 1.0-n;
   return (n);
}
```

1.5.2 *Quantile of Normal Distribution*

```
/*--------------------------------------------------------------*/
/* Remark: Compute the quantile of the standard normal dist */
```

```
/*   parameters: beta between 0 and 1                            */
/*                                                                */
double quantile_normal(double beta)
{
double zz, n, x1, t;

   x1 = beta;

   if (beta < 0.5) beta = 1 - beta;

   t = sqrt(-2 * log(1-beta));

   zz = 2.515517 + t * (0.802853 + 0.010328 * t);

   n = 1 + t * (1.432788 + t * (0.189269 + 0.001308*t));
   zz = t - zz/n;

   if (x1 < 0.5)
      return (-zz);
   else
      return (zz);

}
```

Other procedures to compute quantile function of the standard normal distribution are available, e.g., http://home.online.no/~pjacklam/notes/invnorm/impl/sprouse/ltqnorm.c.

1.5.3 *Quantile of t Distribution*

```
/*------------------------------------------------------------*/
/* Remark: Compute the quantile of the t-distribution          */
/*   parameters: beta between 0 and 1                          */
/*   n is degrees of freedom                                   */
/*                                                             */
double quantile_t(double beta, int n)
{
double t, a, c, u;
```

```
   if (n == 1) return (tan(M_PI*(beta-0.5)));

   if (n == 2) {
      t = 2*beta - 1;
      return (sqrt(2)*t/sqrt(1-t*t));
   }; /* n == 2 */

   a = n - (2.0/3.0) + 1.0/(10*n);
   c = (n - (5.0/6.0)) / (a*a);
   u = quantile_norm(beta);
   t = sqrt(n*exp(c*u*u) - n);

   if (beta > 0.5)
      return (t);
   else
      return (-t);
}
```

1.5.4 *Quantile of Chi-square Distribution*

```
/*----------------------------------------------------------------*/
/* Remark: Compute the quantile of the chi-square dist           */
/*   parameters: beta between 0 and 1                            */
/*   n is degrees of freedom                                     */
/*                                                               */
double quantile_chi(double beta, int n)
{
double chi, h60, q, zz;
int c;
double h[15];

   h[0] =  0.0118;   h[1] = -0.0067;   h[2] = -0.0033;
   h[3] = -0.0010;   h[4] =  0.0001;   h[5] =  0.0006;
   h[6] =  0.0006;   h[7] =  0.0002;   h[8] = -0.0003;
   h[9] = -0.0006;   h[10] = -0.0005;  h[11] =  0.0002;
   h[12] =  0.0017;  h[13] =  0.0043;  h[14] =  0.0082;

   c = -1;
   q = quantile_norm(beta);
```

```
   zz = -3.5;

   while (zz < q) {
      c += 1;
      zz += 0.5;
   }

   if (q < -3.5)
      h60 = 0;
   else
      h60 = h[c]+ 2*(h[c+1]-h[c])*(q-zz+0.5);

   chi = 1 - 2.0/(9*n)+(q-60/n*h60)*sqrt(2.0/(9*n));

   return  (n*exp(3*log(chi)));
}
```

```
/* chi square test */
int chisquare_test(int* r, double* p, int df)
{
int i;
int total = 0;
double score = 0.0;

   for (i = 0; i <= df; ++i)
       total += r[i];

   for (i = 0; i <= df; ++i)
       score += pow(r[i] - total*p[i],2) / (total*p[i]);

   if (score < quantile_chi(0.9, df))
       return (1);
   else
       return (0);
}
```

1.5.5 *Standard Deviation*

Standard deviation can be computed by $S^2(N) = (\sum_1^N X_i - \bar{X})^2/(N-1)$, where $\bar{X} = \sum_1^N X_i/N$. However, if we use the equation $S^2(N) = (\sum_1^N X_i^2/N - \bar{X}^2)N/(N-1)$ to compute the variance estimator, instead of $(X_1, X_2, \ldots, X_N)$, we are only required to store the triplet $(N, \sum_i^N X_i, \sum_i^N X_i^2)$.

It seems straightforward to compute $\bar{X} = \sum_1^N X_i/N$. However, if the sample size N is a very huge number, we may encounter problems. For example, the value of $A = \sum_i^n X_i$ for some $n < N$ may be much greater than X_{n+1}, consequently, the value of $\sum_i^{n+1} X_i$ will equal to A as well. The severity of the error becomes greater when the difference of $N - n$ becomes larger.

```
/*-------------------------------------------------------------*/
/* Remark: Compute the variance: algorithm 1                   */
/*   parameters: n is the number of values in the Array        */
/*   Array contains n values                                   */
/*   mean is the output and is the average of n values         */
/*                                                             */
double variance1(int n, double* Array, double* mean)
{
int i;
double variance=0.0;

   *mean = 0.0;
   for (i = 0; i < n; ++i)
       *mean += Array[i];

   *mean /= n;

   for (i = 0; i < n; ++i)
       variance+=pow(Array[i]-*mean,2);

   return(variance / (n-1));
}
```

```
/*----------------------------------------------------------------*/
/* Remark: Compute the variance: algorithm 2                      */
/*    parameters: n is the number of values used to compute       */
/*                Sumx and SumX2                                  */
/*    SumX is the sum of n values                                 */
/*    SumX2 is the sum of the square of n values                  */
/*                                                                */
double variance2(int n, double SumX, double SumX2)
{
double xAvg = SumX/n;
double x2Avg = SumX2/n;
double variance = (x2Avg - pow(xAvg,2));

   variance = variance * n / (n-1);

   return (variance);
}
```

1.6 Summary

We have discussed the backbone CMRG random-number generator, several utilities to enhance its practical use, and implementation issues. This implementation eliminates the need to use a fixed number of pre-computed seeds, which provides little flexibility. The presented random-number-generation package provides jumping facilities, has good speed, a long period length, and excellent theoretical/statistical properties. It provides for multiple generators (streams) running simultaneously, and each generator (stream) has its sequence of numbers partitioned into many long disjoint contiguous substreams. The RNG mrg32k3a has been implemented in several software packages. We also provided some subroutines to generate non-uniform random variables and some useful utilities.

Chapter 2

Sample Sizes and Stopping Rules

Simulation can be generally classified into two types: 1) Finite-Horizon (terminating) simulation: a simulation where there is a specific starting and stopping condition that is part of the model; and 2) Steady-State (non-terminating) simulation: a simulation where there is no specific starting and ending conditions. The purpose of a steady-state simulation is the study of the long-run behavior of a system. A performance measure is called a steady-state parameter if it is a characteristic of the equilibrium distribution of an output stochastic process. Steady-state distribution does not depend on initial conditions (when the initialization bias have been properly taken care of), but the nature and rate of convergence of the transient distributions can depend heavily on the initial conditions. The length of the simulation must be determined dynamically to ensure that a representative steady-state model behavior is reached. We seek algorithms that provide good results and do not affect the performance of a given system.

Stopping rules are one of the crucial components of sequential simulation procedures affecting the performance of such procedures. These rules are often used to determine the simulation run length (sample size). Many stopping rules aim to provide an accurate estimate of the sampling error of a certain unknown parameter. The rule would give the experimenter an idea of the precision with which the estimate reflects the true but unknown parameter. For example, when estimating a steady-state performance parameter such as the mean μ of some discrete-time stochastic output process $\{X_i : i \geq 1\}$ via simulation, it is desirable to devise an algorithm to determine the simulation run length n so that 1) the mean estimator (i.e., the sample mean $\bar{X}(n) = \sum_{i=1}^{n} X_i/n$) is unbiased; 2) the confidence interval (c.i.) for μ is of a prescribed width; and 3) the actual coverage probability of the c.i. is close to the nominal coverage probability $1 - \alpha$. Simulation can also be classified into two categories: 1) discrete (time) event; and 2) continuous time. We focus on discrete-event simulation in our discussion.

Before a simulation can be run, one must provide initial values for all of the simulation's state variables. Since the experimenter may not know what initial values are appropriate for the state variables, these values might be chosen somewhat arbitrarily. For instance, we might decide that it is "most convenient" to initialize a queue as empty and idle. Such a choice of initial conditions can have a significant but unrecognized impact on the simulation run's outcome. It is well known that just after initialization (presumably the initialized state is not steady-state) any stochastic system with random inputs is in a transient state, during which its stochastic characteristics vary with time and differ from the steady-state distribution. This is caused by the fact that the stochastic system initially moves along nonstationary paths. After a period of time, the system approaches its statistical equilibrium on a stationary path if the system is stable, or remains permanently on a nonstationary path if the system is not stable [Pawlikowski (1990)].

Output data collected during transient periods do not characterize steady-state behavior of simulated systems, so they can cause quite significant bias of the final steady-state results. The influence that the initial transient data can have on the final results is a function of the strength of the autocorrelation of collected observations, the length of a simulation run, and the rate of convergence to steady state. Thus, the initialization bias problem can lead to errors, particularly in steady-state output analysis.

The mean estimator would be unbiased provided that the underlying process is stationary; i.e., the joint distribution of the X_i's is insensitive to time shifts. That is, the simulation could theoretically go on indefinitely with no statistical change in behavior. In cases where the process is not strictly stationary due to initialization effects, this influence can be arbitrarily weakened by running the simulating program sufficiently long. But in practical situations simulation experiments are restricted in time and it is not known in advance what is the required simulation run length to eliminate the effect of initial transient. One can usually resort to approximate stationarity by removing some number of initial observations from sequential simulation. Determining the number of initial observations to be removed is itself a research topic and we do not address this issue here. Moreover, many procedures have been developed to determine the initial transient period, see [Hoad et al. (2010)]. Those procedures can be applied to determine the warm-up period before using procedures that require the underlying process to be stationary.

Simulation results do not provide exact answers, those results are only estimates. Hence, statistical methods are required to analyze the results of simulation experiments. The usual method of c.i. construction from classical statistics, which requires i.i.d. normal observations, is not directly applicable since simulation output data are generally correlated and non-normal. Statistical analysis needs to be performed to estimate the standard error and confidence interval as well as to figure out the number of observations required to achieve a desired error or confidence interval. Given that the sample size and consequently the quality of c.i. are largely determined by the stopping rules, there is significant interest in further expanding knowledge base of stopping rules for sequential simulation procedures. There exists many stopping rules, see [Law (2014); Singham (2014)] for details. Many of these existing techniques can be either overly restrictive to be applied to general cases or excessively complicated to be implemented for practical use.

We discuss a generic stopping rule to determine simulation run length by test of independence, which can be easily incorporated into simulation procedures. The experimental results indicate this stopping rule works well. These procedures perform statistical analysis on systematic samples that are collected on strictly stationary stochastic processes. The systematic samples are obtained by *systematic sampling*; i.e., first select a number l between 1 and L (a positive integer), choose that observation and then every l^{th} observation thereafter. Here, the chosen l will be sufficiently large so that observations l apart appear to be independent. We need to specify v, the required number of near-independent observations, and l for the systematic sampling. The minimum required total run length is then $N = vl$.

The procedures sequentially increase the lag l, and consequently the simulation run length, until a sequence of (v) systematic samples passes the test of independence. The simulation run length thus derived is based entirely on the data proper and requires no user intervention, provided that the autocorrelation of the stochastic-process output sequence dies off as the lag between observations increases, in the sense of ϕ-mixing, see Section 2.1. These mild assumptions are satisfied in virtually all practical settings. Roughly speaking, a stochastic process is ϕ-mixing if its distant future behavior is essentially independent of its present or past behavior [Billingsley (1999)].

We investigate two procedures of test of independence: 1) the von Numann [von Neumann (1941)] test; 2) runs test. One drawback of using the runs test to determine whether a sequence appears to be independent is that it can require a large sample size, for example, 4000 or more samples. This trait may render the proposed stopping rule cost-prohibitive for certain instances. As an alternative, one may employ the von Neumann test, which requires a much smaller sample size; for example, 180 samples. Note that von Neumann test is computationally more expensive and less powerful than the runs test. Hence the alternative proposed herein represents a tradeoff option that the experimenter can make when constructing the simulation procedures. The proposed stopping rule can be used as a generic method to determine the simulation run length for many simulation procedures. Based on the nature of the underlying simulation study, different tests of independence can be used to determine the simulation run length. As an illustration, we apply this stopping rule with the von Neumann test of independence to determine the batch size of batch means, see [Chen (2012)].

2.1 Definitions

This section reviews 1) ϕ-mixing; 2) the batch-means method, which is used to estimate the variance of the sample mean; and 3) the von Neumann test and the runs test, which determine whether a sequence is sufficiently independent for the purpose of simulation tasks.

Let $\{X_i; -\infty < i < \infty\}$ be a stationary sequence of random variables defined on a probability space (Ω, F, P). Thus, if $M_{-\infty}^k$ and M_{k+j}^∞ are, respectively, the σ-fields generated by $\{X_i; i \le k\}$ and $\{X_i; i \ge k+j\}$, and if $E_1 \in M_{-\infty}^k$ and $E_2 \in M_{k+j}^\infty$, then for all $k(-\infty < k < \infty)$ and $j(j \ge 1)$, if

$$|\Pr(E_2|E_1) - \Pr(E_2)| \le \phi(j),$$

where $1 \ge \phi(1) \ge \phi(2) \ge \cdots$, and $\lim_{j\to\infty} \phi(j) = 0$, then $\{X_i; -\infty < i < \infty\}$ is called *ϕ-mixing*. For more information on ϕ-mixing, see [Billingsley (1999)]. Intuitively, $X_1, X_2, \cdots, X_n$ is ϕ-mixing if X_i and X_{i+j} become virtually independent as j becomes large.

These weakly dependent, stationary processes typically obey a Central Limit Theorem (CLT) of the form

$$\sqrt{n}\frac{[\bar{X}(n) - \mu]}{\Omega} \xrightarrow{D} N(0,1) \text{ as } n \to \infty, \tag{2.1}$$

where Ω^2 is the steady-state variance constant (SSVC), $N(\mu, \sigma^2)$ denotes the normal distribution with mean μ and the variance σ^2, and $\xrightarrow{D}$ denotes convergence in distribution [Billingsley (1999)].

For correlated sequences, the SSVC

$$\Omega^2 \equiv \lim_{n\to\infty} n\mathrm{Var}[\bar{X}(n)] = \sum_{i=-\infty}^{\infty} \gamma_i,$$

where $\gamma_i = \mathrm{Cov}(X_k, X_{k+i})$ for any k is the lag-i covariance. A sufficient condition for the SSVC to exist is that the output process is stationary and $\sum_{-\infty}^{\infty} |\gamma_i| < \infty$. If the sequence is independent, then the SSVC is equal to the process variance $\sigma_x^2 = \mathrm{Var}(X_i)$. For a finite sample n, let

$$\sigma^2(n) = \gamma_0 + 2\sum_{i=1}^{n-1}(1 - i/n)\gamma_i.$$

It follows that $\lim_{n\to\infty} \sigma^2(n) = \sigma^2$ so

$$\mathrm{Var}[\bar{X}(n)] = \sigma^2(n)/n \approx \sigma^2/n,$$

provided that n is sufficiently large. The procedures may fail when the underlying stochastic process does not satisfy the limiting result of Eq. (2.1).

Furthermore, for ϕ-mixing sequences, the natural estimators (see Section 6.1.1) of other distribution characteristics also perform well. For $0 < p < 1$, the pth quantile (percentile) of a distribution is the value at or below which $100p$ percent of the distribution mass lies. Quantile estimation can be computed using standard non-parametric estimation based on order statistics, which can be used not only when the data are i.i.d. but also when the data are drawn from a stationary, ϕ-mixing process of continuous random variables. It is shown in [Sen (1972)] that quantile estimates, based on order statistics, have a normal limiting distribution and are asymptotically unbiased, if certain conditions are satisfied.

Although asymptotic results are often applicable when the amount of data is "large enough," the point at which the asymptotic results become valid generally depends on unknown factors. An important practical decision must be made regarding the sample size n required to achieve the desired precision. Therefore, both asymptotic theory and workable finite-sample approaches are needed by the practitioner.

2.2 Batch-Means Method

The Batch-Means (BM) method is a well-known technique for estimating the variance of point estimators computed from simulation experiments. It attempts to reduce autocorrelation by batching observations. In the non-overlapping BM (NOBM) method, the simulation output sequence $\{X_i : i = 1, 2, \ldots, n\}$ is divided into b adjacent non-overlapping batches, each batch of a size m. The sample size n is therefore $n = bm$. The sample mean, $\bar{X}_j$, for the jth batch is

$$\bar{X}_j = \frac{1}{m} \sum_{i=m(j-1)+1}^{mj} X_i \quad \text{for } j = 1, 2, \ldots, b.$$

For the given batch size m, we have

$$\mathrm{Var}[\bar{X}_j] = \Omega^2(m)/m, \text{ where } \Omega^2(m) = \gamma_0 + 2\sum_{i=1}^{m-1}(1 - i/m)\gamma_i.$$

The grand mean $\hat{\mu}$ of the individual batch means, given by

$$\hat{\mu} = \frac{1}{b}\sum_{j=1}^{b} \bar{X}_j, \tag{2.2}$$

is used as a point estimator for μ. Here $\hat{\mu} = \bar{X}(n) = \sum_i^n X_i/n$, the sample mean of all n individual X_i's, and we seek to construct a c.i. for μ based on the point estimator of Eq. (2.2).

The BM variance estimator (for estimating $\mathrm{Var}[\bar{X}_j]$) is simply the sample variance of the mean estimator $\bar{X}_j$ computed from the batch means

$$S^2 = \frac{1}{b-1}\sum_{j=1}^{b}(\bar{X}_j - \hat{\mu})^2. \tag{2.3}$$

Consequently, the batch-means estimator for Ω^2 is $\hat{\Omega}_B^2 \equiv mS^2$.

Asymptotic validity of the c.i. constructed by the batch-means method - i.e., the coverage probability of the c.i. is close to the nominal coverage probability - tend to depend on the validity of the assumption that the batch means are approximately i.i.d. normal. That is, for a large batch size m, the batch means are approximately i.i.d. normal with unknown mean μ and unknown variance $\Omega^2(m)/m$. Other BM methods, e.g., [Tafazzoli et al. (2011)], may choose to construct a c.i. for the mean by adjusting the c.i. half-width based on the strength of the autocorrelations between BM as well as the skewness of BM. There are many BM methods, see [Hoad et al. (2011)] for the 'family trees' of BM methods in the simulation literature.

2.3 Determining the Simulation Run Length

In this section, we discuss the von Neumann test of independence, the runs test of independence, the strategy of using test of independence to implement a stopping rule as well as a procedure to construct a confidence interval for the mean.

2.3.1 *The von Neumann Test of Independence*

The von Neumann test of independence is relatively simple and most adequate for cases calling for small sample sizes. The test constructs a metric commonly known as the von Neumann ratio. Several BM procedures have used the von Neumann ratio to test for independence, e.g., [Fishman (1978)].

The von Neumann test [von Neumann (1941); Fishman (2001)] can be put forth as the following. For the null hypothesis H_0: the samples $X_1, X_2, \ldots, X_v$ are uncorrelated, the von Neumann ratio is

$$C_v = 1 - \frac{\sum_{i=2}^{v}(X_i - X_{i-1})^2}{2\sum_{i=1}^{v}(X_i - \bar{X}(v))^2}.$$

Note that C_v is an estimator of the lag-1 autocorrelation $\omega_1 \equiv \text{Corr}(X_i, X_{i+1})$, adjusted for end effects that diminish in importance as the number of samples v increases. Note that v is the sample size relevant to the von Neumann test and is generally different than the simulation run length n.

The von Neumann test statistic for H_0 is

$$Z = \sqrt{\frac{v^2 - 1}{v - 2}}C_v.$$

Under H_0, $Z \sim N(0, 1)$, so one rejects H_0 at level $1 - \alpha_{\text{ind}}$ if $Z > z_{1-\alpha_{\text{ind}}}$, where $z_{1-\alpha_{\text{ind}}}$ is the $1 - \alpha_{\text{ind}}$ quantile of the standard normal distribution.

[Fishman (2001)] points out that the von Neumann test of independence is likely to accept H_0 when $\{X_i\}$ has an autocorrelation function that is negatively correlated and exhibits damped harmonic behavior around zero. In this case, the variance estimator is biased high instead of biased low. While one does not lose any performance in terms of coverage, the half-width of the confidence interval for μ will be wider than otherwise achievable.

2.3.2 *A Source Code of the von Neumann Test*

This subroutine implements the von Neumann test of independence.

```
/* von Neumann test of independence */
int vonNeumann(double* BatchN, int bnum, double muhat)
{
double t;
int i;
double temp1 = 0.0;
double temp2 = 0.0;

    temp2 = pow(BatchN[0] - muhat,2);

    for (i = 1; i < bnum; ++i) {
        temp1 += pow(BatchN[i] - BatchN[i-1],2);
        temp2 += pow(BatchN[i] - muhat,2);
    };

    t = sqrt((bnum*bnum-1)/(bnum-2))*(1-temp1/(2*temp2));

    if ( t <= quantile_norm(0.9) ) {
          return (1);
    } else {
          return (0);
    };
}
```

2.3.3 *The Runs Test of Independence*

Another test of independence is the runs test, it examines the sequence of output data for unbroken subsequences of maximal length within which the sequence increases (decreases) monotonically; such a subsequence is called a *run up* (*run down*), and the length of the subsequence is called the *run length*. Note that this run length (for the test of independent) is different from the simulation run length. The run length used by the runs test is to determine whether a sequence appears to be independent. On the other hand, simulation run length is the total length of a particular simulation execution, measured either in number of observations collected (for a discrete-time output process), or amount of simulated time (for a continuous-time output process); we focus on discrete-time processes. The runs test looks solely for lack of independence and has been shown to be very

```
/* check whether the sequence is independent */
#RUNSUPSZ   4000
int checkind(double* Observ, double seed[6])
{
/* proportion of each run length */
static double p[6] = {1/2,1/3,1/8,1/30,1/144,1/840};
int i;
int r[6] = {0, 0, 0, 0, 0, 0};;
int runup=1;
int df = 5;

   for (i = 1;  i < RUNSUPSZ; ++i ) {
   /* save run up statistics */
      if (runup == 0) {
          runup = 1;
      } else if (Observ[i] > Observ[i - 1]) {
         ++runup;
      } else {
         if ((Observ[i] == Observ[i - 1]) &&
             (RnadU01(seed) < 1.0/(runup+1))) {
            ++runup;
         } else {
            if (runup >= df+1)
               ++r[df];
            else
               ++r[runup-1];

            runup = 0;
         };
      };
   }; /* for i */

   if ( chisquare_test(r, p, df) )
      return (1);
   else
      return (0);
}
```

2.3.5 *An Implementation of Determining the Simulation Run Length*

Sequential procedures usually start a simulation run with an initial sample size, which may be determined in advance or dynamically at run time. Based on information obtained with the initial samples, sequential procedures then progressively increase the sample size until the prescribed stopping rule is satisfied. If the estimator satisfies the specified precision, then the procedure returns the estimated value and terminates. Otherwise, the procedure continues the simulation, either by adding more samples in the current replication or more replications.

Let l_i be the lag at which systematic samples pass the test of independence. We call the sequence of the (lag-l_i) systematic samples that appear to be independent the *quasi-independent* (QI) sequence. Note that independent sequences are likely to pass the test of independence with lag 1 observations. Hence, the procedure will allocate small sample sizes when the underlying output sequences are independent. This allows users to use the QI procedures without allocating excessively large number of samples. Furthermore, the QI stopping rule can be used with various sequential procedures to estimate mean, variance, quantile, density, proportion, etc.

In the following section, we apply a BM method to estimate the SSVC and construct a c.i. of the steady-state mean employing the proposed stopping rules as an illustration.

2.4 Constructing the Confidence Interval

In the approach, we first compute the lag l for the systematic samples. Once the lag l is large enough for the lag-l_s systematic samples in the buffer to pass the test of independence, we check whether the intermediate batches (i.e., the averages of l_s consecutive primitive batches that will be discussed later) appear to be independent. Note that the lag at which systematic samples appear to be independent $l_i = ll_s$. When the intermediate batches appear to be independent, we then compute the batch means (by taking the average of several consecutive intermediate batches) and the variance estimator S^2 based on the batch means. Otherwise, the procedure will continue to increase the lag l_i and consequently the sample size.

The final step of the procedure is to determine whether the c.i. meets the user's half-width requirement. This requirement usually takes one of the following forms: 1) a maximum absolute half-width ϵ; or 2) a maximum relative fraction γ of the magnitude of the final mean estimator $\hat{\mu}$. Let $w = t_{b-1,1-\alpha/2}S/\sqrt{b}$ denote the c.i. half-width. Here $t_{f,1-\alpha}$ is the $1-\alpha$ quantile of the t distribution with f degrees of freedom. If the precision of the c.i. is satisfied (i.e., $w \le \epsilon$ or $w \le \gamma|\hat{\mu}|$), then the procedure terminates, and we return the mean estimate $\hat{\mu}$ and the c.i. with half-width w. If the precision requirement is not satisfied, the procedure will increase the sample size to shorten the half-width. The procedure will increase the number of batches b to

$$(w/\epsilon)^2 b \text{ or } (w/(\gamma\hat{\mu}))^2 b. \tag{2.4}$$

This step will be executed repeatedly until the half-width is within the specified precision.

2.5 A Correlation Adjustment

There may be residual correlations between the final batch means, which will cause the c.i. half-width to be narrower than necessary and result in coverages less than the specified nominal value. The strategy of SBatch [Lada et al. (2008)] can be used to adjust the c.i half-width for any residual correlation between the batch means. The correlation-adjusted c.i. half width

$$w_A = \sqrt{A}w. \tag{2.5}$$

The correlation adjustment A is computed as

$$A = \frac{1 + \hat{C}_{\bar{X}}}{1 - \hat{C}_{\bar{X}}},$$

where the standard estimator of the lag-1 correlation of the batch means is

$$\hat{C}_{\bar{X}} = \widehat{\mathrm{Corr}}(\bar{X}_j, \bar{X}_{j+1}) = \frac{1}{b}\sum_{j=1}^{b-1}(X_j - \hat{\mu})(X_{j+1} - \hat{\mu})/S^2.$$

Note that in the implementation we set $A = 1$ when $A < 1$ so that $w \le w_A$.

2.6 An Implementation of Batch-Means Method

We use the QI stopping rule to implement a batch-means procedure, QI-Batch2. Figure 2.1 displays a high-level flow chart of QIBatch2. We allocate two buffers with size $3v$ to keep systematic samples and *primitive batches* (PB), i.e., the average of l observations. Batch means are then computed from PB in the buffer. Initially $l = 1$ and each observation is treated as a PB; as the procedure proceeds, l will be doubled every two sub-iterations. The steps between generating more observations are considered a sub-iteration.

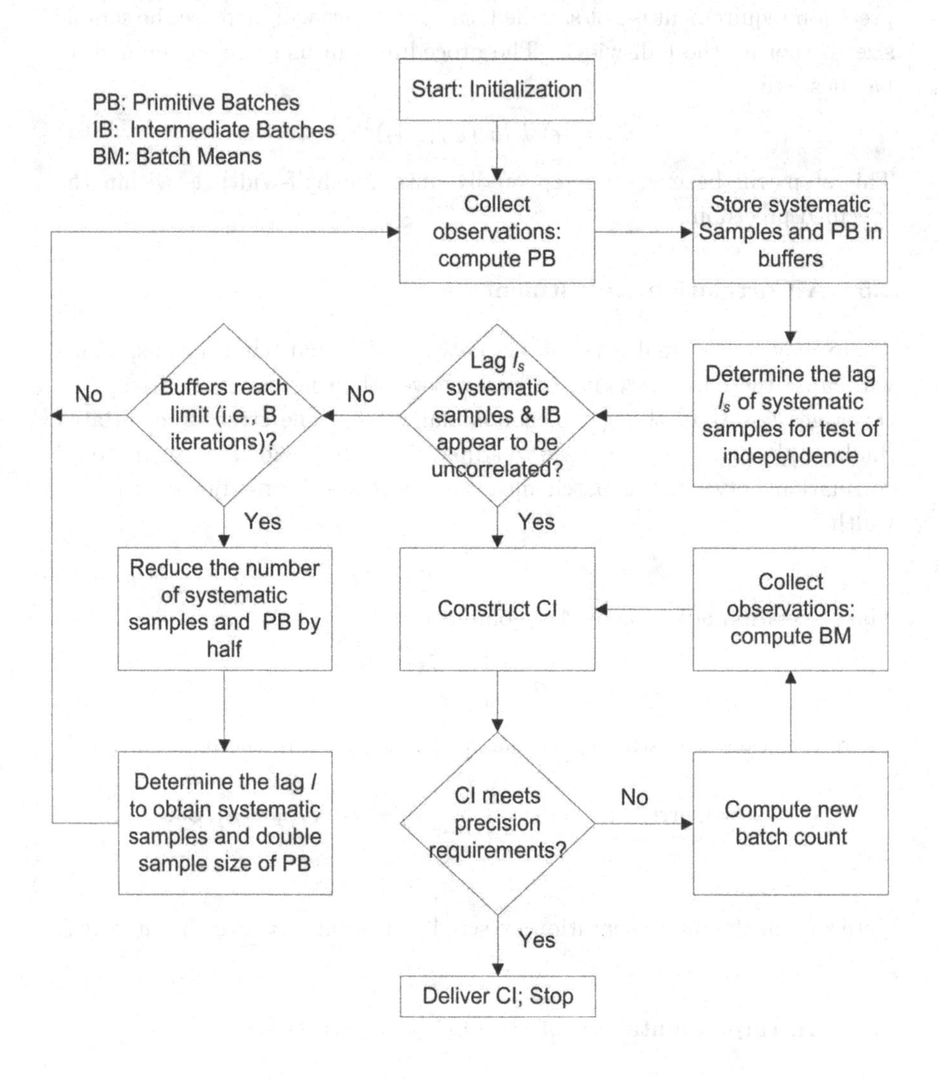

Fig. 2.1 High-level flow chart of QIBatch2

To facilitate the description of the algorithm, we index the sub-iterations with subscript A and B so that most relevant quantities can be described in simpler forms. The procedure performs the test of independence in both A and B sub-iterations and terminates when both the lag-l_s systematic samples in the buffer and the *intermediate batches* (IBs, the average of l_s adjacent PBs) appear to be independent.

Finally, we aggregate the available PBs into 30 BMs by averaging the adjacent PBs. Using 30 as the final number of batches is somewhat arbitrary. The rationale is to insure that the number of batches is large enough to obtain a good autocorrelation estimate.

The QIBatch2 algorithm:

The size of the buffer used to store the PB is $3v$, l is the lag used to create the systematic samples and is also the number of observations used to compute the PB, δ is the incremental sample size, r is the index of iterations. Each iteration r contains two sub-iterations r_A and r_B.

(1) Initialization: Set $v = 180$, $b = 30$, $q = v/b$, $l = 1$, $l_s = 1$, $\delta = v$, and $r = 0$.
(2) Generate δ systematic samples and PBs, where each PB is the average of l observations.
(3) If this is a r_A ($r > 0$) iteration, set $l_s = 2$. If this is a r_B ($r > 0$) iteration, set $l_s = 3$.
(4) Compute the intermediate batches, i,e., the average of l_s adjacent PBs.
(5) Check whether the lag-l_s systematic samples in the buffer and the intermediate batches pass the von Neumann test.
(6) If these systematic samples and intermediate batches pass the von Neumann test, go to step 11.
(7) If this is the initial or a r_B iteration, set $r = r + 1$ and start a r_A iteration. If this is a r_A iteration, start a r_B iteration.
(8) If this is a r_A ($r > 1$) iteration, then re-calculate the PBs in the buffer by taking the average of two consecutive PBs, and reindex the rest of the $3v/2$ PBs in the first half of the buffer. Re-index the systematic samples in the buffer by discarding every other samples, and reindex the rest of the $3v/2$ systematic samples in the first half of the buffer. Set $l = 2^{r-1}$, $\delta = v/2$.
(9) If this is a r_B ($r > 1$) iteration, set $\delta = v$.
(10) Go to step 2.

Table 2.1 Properties of QIBatch2 at each iteration

Iteration	0	1_A	1_B	2_A	2_B	...	r_A	r_B
n	v	$2v$	$3v$	$4v$	$6v$	...	$2^r v$	$2^{r-1}3v$
V_i	v	$2v$	$3v$	$2v$	$3v$	...	$2v$	$3v$
l	1	1	1	2	2	...	2^{r-1}	2^{r-1}
l_i	1	2	3	4	6	...	$2^{r-1}2$	$2^{r-1}3$

(11) Set the value of the batch means to the average of $q'(=ql_s)$ PBs in the buffer (i.e., $\bar{X}_j = \sum_{q'(j-1)+1}^{q'(j-1)+q'} V_j/q'$ for $j = 1, 2, \ldots, b$, where V_j is the value of the jth PB). Compute the batch-means variance estimator S^2 according to Eq. (2.3) and confidence interval half-width according to Eq. (2.5).

(12) Let ϵ be the desired absolute half-width, and let $\gamma|\hat{\mu}|$ be the desired relative half-width. If the half-width of the c.i. is greater than ϵ or $\gamma|\hat{\mu}|$, compute b', the required number of batches according to Eq. (2.4), generate $\delta' = \lceil (b'-b)/2 \rceil$ additional batches, set $b = b + \delta'$, and go to step 11; otherwise the procedure returns the c.i. estimator and terminates.

The QIBatch2 procedure starts with an initial sample size of 180 and doubles the sample sizes every two sub-iterations. We choose the value $v = 180$ because this is the sample size we used for the von Neumann test. Note that we set $\delta = v/2$ at A iterations and $\delta = v$ at B iterations so that the number of PBs will always be a multiple of v at the end of an iteration. The procedure progressively increases the lag l until the lag-l_s systematic samples in the buffer and the intermediate batches pass the von Neumann test. The procedure needs only to process each observation once and does not require storing the entire output sequence. See Section 2.7 for some properties of the procedure as it proceeds from iteration to iteration.

We calculate the required number of batch to meet the specified precision requirement by Eq. (2.4) when the default precision does not meet the requirement. However, instead of allocating the entire additional number of batches, we allocate only half of those, i.e., $\delta' = \lceil (b'-b)/2 \rceil$ at each iteration. This is to reduce the frequency of over allocating the required number of batches. The mean μ and the c.i. half-width are estimated, respectively, by $\hat{\mu}$ and Eq. (2.5).

2.7 An Illustration of Allocated Sample Sizes

Table 2.1 illustrates how sampling progresses from iteration to iteration. The *Iteration* row lists the running index of the iterations. The n row lists the total number of observations to be taken during a certain iteration. The V_i row lists the total number of PBs in the buffer at a certain iteration. The l row lists the lag l that is used to obtain the systematic samples in the buffer. Note that l is also the number of observations used to obtain the PB in the buffer. The l_i row lists the lag l_i at which the systematic samples appear to be independent when the lag l_s of the systematic samples in the buffer pass the von Neumman test of independence.

For example, at the end of sub-iteration 1_B, the total number of observations is $3v$, there are $3v$ PBs in the buffer, and each PB is just the value of each observation. At the beginning of sub-iteration 2_A, we reduce the number of PBs in the buffer from $3v$ to $3v/2$ by taking the average of every two consecutive PBs. By reducing the number of PBs in the buffer at $r_A(r > 1)$ iterations, the number of PBs will always be no more than $3v$. We will generate $v/2$ PBs (with the new increased batch size 2) at sub-iteration 2_A; so we will have $2v$ PBs at the end of the iteration.

Table 2.2 lists the value of the PBs times 2^{r-1} at each iteration. Note that the size of number of observations of each PB at the r iterations is 2^{r-1}. That is, the value of each PB (i.e., V_i for $i = 1, 2, \ldots, 3v$) is the average of 2^{r-1} observations. Here x_i is the realization of X_i. At the end of the initial iteration, there will be v PBs. At the end of the r_A and r_B iterations, there will be $2v$ and $3v$ PBs, respectively. Let ϱ_i for $i = 1, 2, \ldots, v$ denote the ith intermediate batch. Table 2.3 lists the value of the v intermediate batches that are used for the von Neumann test of independent. Once these v intermediate batch means appear to be independent, the $b = 30$ final batch means are obtained by averaging the consecutive $q = v/b$ intermediate batch means. Table 2.4 lists the value of the final b batch means.

2.8 Empirical Experiments

In this section, we present some empirical results from simulation experiments using the proposed procedure. We test the procedure with six stochastic processes:

- Observations are i.i.d. uniform between 0 and 1, denoted $U(0, 1)$.
- Observations are i.i.d. $N(0, 1)$.

Table 2.2 The value of the PBs times 2^{r-1} at each iteration

Iter	V_1	...	V_v	...	V_{2v}	...	V_{3v}
0	x_1	...	x_v				
1_A	x_1	...	x_v	...	x_{2v}		
1_B	x_1	...	x_v	...	x_{2v}	...	x_{3v}
2_A	$\sum_{i=1}^{2} x_i$	...	$\sum_{i=2v-1}^{2v} x_i$	...	$\sum_{i=4v-1}^{4v} x_i$		
2_B	$\sum_{i=1}^{2} x_i$	...	$\sum_{i=2v-1}^{2v} x_i$	...	$\sum_{i=4v-1}^{4v} x_i$	...	$\sum_{i=6v-1}^{6v} x_i$
$\vdots$				$\vdots$			
r_A	$\sum_{i=1}^{2^{r-1}} x_i$	...	$\sum_{i=2^{r-1}(v-1)+1}^{2^{r-1}v} x_i$	...	$\sum_{i=2^r(v-1)+1}^{2^{r-1}2v} x_i$		
r_B	$\sum_{i=1}^{2^{r-1}} x_i$	...	$\sum_{i=2^{r-1}(v-1)+1}^{2^{r-1}v} x_i$	...	$\sum_{i=2^{r-1}(2v-1)+1}^{2^{r-1}2v} x_i$	...	$\sum_{i=2^{r-1}(3v-1)+1}^{2^{r-1}3v} x_i$

Table 2.3 The value of the v intermediate batches that used for the von Neumann test

Iteration	ϱ_1	ϱ_2	...	ϱ_v
0	V_1	V_2	...	V_v
r_A	$(V_1 + V_2)/2$	$(V_3 + V_4)/2$	...	$(V_{2v-1} + V_{2v})/2$
r_B	$(V_1 + V_2 + V_3)/3$	$(V_4 + V_5 + V_6)/3$	...	$(V_{3v-2} + V_{3v-1} + V_{3v})/3$

Table 2.4 The value of the final b batches

$\bar{X}_1$	$\bar{X}_2$	...	$\bar{X}_b$
$\sum_{i=1}^{q} \varrho_i/q$	$\sum_{i=q+1}^{2q} \varrho_i/q$	...	$\sum_{i=(b-1)q+1}^{v} \varrho_i/q$

- Observations are i.i.d. exponential with mean 1, denoted expon(1).
- Steady-state *first-order moving average* process, generated by the recurrence

$$X_i = \mu + \epsilon_i + \theta\epsilon_{i-1} \quad \text{for} \quad i = 1, 2, \ldots,$$

  where the ϵ_i are i.i.d. $N(0,1)$, and $-1 < \theta < 1$. We set μ to 2 in our experiments. This process is denoted MA1(θ).
- Steady-state *first-order autoregressive* process, generated by the recurrence

$$X_i = \mu + \varphi(X_{i-1} - \mu) + \epsilon_i \text{ for } i = 1, 2, \ldots,$$

  where the ϵ_i are i.i.d. $N(0,1)$, and $-1 < \varphi < 1$. We set μ is set to 2 in our experiments. This process is denoted AR1(φ). We set X_0 to a random variate drawn from a $N(0, \frac{1}{1-\varphi^2})$ distribution.
- Steady-state M/M/1 delay-in-queue process with arrival rate λ and service rate $\nu = 1$. This process is denoted MM1(ρ), where $\rho = \lambda/\nu$ is the traffic intensity.

First, we evaluate the performance of the von Neumann test with different levels of α. Based on the experimental results, the confidence level of the von Neumann test is set to 90% in latter experiments. Note that a lower confidence level of these tests will increase the chance of committing a Type I error (rejecting the null hypothesis when it is true) and will increase the simulation run length.

We check the interdependence between the lag l at which the systematic samples pass the independence test and the strength of the autocorrelation of the output sequence. We evaluate the performance of the QIBatch2 procedure to estimate the variance of sample means.

Table 2.5 The percentage of the output sequences pass the von Neumann test of independence with 180 samples at different levels of α

Process	$P(90\%)$	$P(95\%)$	$P(99\%)$
$U(0,1)$	90%	95%	99%
$N(0,1)$	90%	95%	99%
expon(1)	90%	95%	99%
MA1(0.15)	25%	39%	65%
MA1(0.25)	2.5%	6.0%	20%
MA1(0.35)	0.05%	0.23%	1.9%
MA1(0.50)	0%	0%	0.02%
MA1(0.75)	0%	0%	0%
AR1(0.15)	24%	37%	63%
AR1(0.25)	2.1%	4.9%	16%
AR1(0.35)	0.03%	0.1%	0.85%
AR1(0.50)	0%	0%	0.01%
AR1(0.75)	0%	0%	0%
MM1(0.15)	23%	28%	38%
MM1(0.25)	3.7%	5.2%	9.4%
MM1(0.35)	0.28%	0.46%	1.1%
MM1(0.50)	0%	0.01%	0.02%
MM1(0.75)	0%	0%	0%

2.8.1 *Experiment 1*

In this experiment, we used the von Neumann test to check whether a sequence of observations passes the tests of independence. The number of observations used to perform the test is $v = 180$. Table 2.5 lists the experimental results. Each design point is based on 10,000 independent simulation runs. The $P(100(1-\alpha_{\text{ind}})\%)$ columns list the observed percentage of these 10,000 runs passing the von Neumann test when the nominal probability of the test of independence is set to $1 - \alpha_{\text{ind}}$. We set α_{ind} and α_{nor} to 0.1, 0.05, and 0.01.

The von Neumann test performs very well in terms of Type I error (i.e., independent sequences fail the test of independence) – it is close to the specified α_{ind} level. A Type II error is the event that we accept the null hypothesis when it is false, i.e., correlated sequences pass the test of independence. For slightly correlated sequences, the frequency of committing a Type II error is high, for example, the MA1(0.15), AR1(0.15), and MM1(0.15) processes. Note that we can use a larger number of samples to increase the power of a test and reduce the Type II error. If we mistakenly treat slightly correlated sequences as being i.i.d., the performance measurements should still be fairly accurate; thus, the low probability of correct decision for those slightly correlated sequences should not pose a

Table 2.6 Average lag l at which the output sequences pass the test of independence

Process	MA1(θ)		AR1(φ)		MM1(ρ)	
θ,φ,ρ	$\bar{l}$	$stdv(l)$	$\bar{l}$	$stdv(l)$	$\bar{l}$	$stdv(l)$
0.15	1.84	0.58	1.88	0.62	2.13	0.86
0.25	2.08	0.37	2.36	0.63	3.09	1.09
0.35	2.11	0.34	2.82	0.74	4.44	1.42
0.50	2.10	0.33	3.85	0.87	7.72	2.14
0.75	2.12	0.36	7.85	1.39	28.24	6.31
0.90	2.11	0.34	18.28	2.65	138.16	27.84

problem. On the other hand, for mildly or highly correlated sequences, the von Neumann test detects the dependence almost all the time.

The experimental results of the runs test of independence are available in [Chen and Kelton (2003)].

2.8.2 *Experiment 2*

In this experiment, we check the interdependence between the average lag at which the systematic samples appear to be independent and the strength of the autocorrelation of the output sequence. We set $\alpha_1 = 0.1$. Table 2.6 lists the experimental results. Each design point is based on 1,000 independent simulation runs. The θ, φ, ρ column lists the coefficient values of the corresponding stochastic process. The MA1(θ) column lists the results of the underlying moving average output sequences. The $\bar{l}$ column lists the average lag at which the systematic samples appear to be independent. Hence, the average computation run length for that particular stochastic process is $v\bar{l}$. The $stdv(l)$ column lists the standard deviation of the lag l. In general, the average lag at which the systematic samples appear to be independent increases as the autocorrelation increases. The MA1(θ) processes are only slightly correlated even with θ as large as 0.9. Therefore, lag 2 observations of the MA1(0.90) output sequences generally appear to be independent. On the other hand, the MM1(0.90) output sequences are highly correlated, the average lag at which the systematic samples appear to be independent is as large as 138. The results from this experiment indicate a strong correlation between the strength of the autocorrelation and the average lag l at which the systematic samples appear to be independent. We also performed the experiment using different v. The average lag at which the systematic samples appear to be independent generally increases with a larger v.

The experimental results of the runs test of independence are available

Table 2.7 Coverage of 90% confidence intervals of independent samples

Process	$U(0,1)$	$N(0,1)$	expon(1)
μ	0.50	0.00	1.00
avg samp	202	203	204
avg hw	0.042	0.142	0.144
stdv hw	0.015	0.049	0.054
coverage	90.7%	90.4%	90.6%

Table 2.8 Coverage of 90% confidence intervals of auto-correlated samples

Process	MA1(0.9)	AR1(0.9)	MM1(0.9)
μ	2.00	2.00	9.00
avg samp	416	13928	243671
avg hw	0.191	0.182	0.814
stdv hw	0.068	0.074	0.397
coverage	90.3%	92.4%	87.4%

in [Chen and Kelton (2003)].

2.8.3 *Experiment 3*

In this experiment, we use the QIBatch2 procedure to construct c.i. of means. Since batch means are often used to estimate the variance of sample means, we evaluate the accuracy of the variance estimated by the procedure. In these experiments, no relative or absolute precisions were specified, so the half-width of the c.i. is the result of the default precision. In this experiment we list the results of c.i. coverage when the entire sequences are divided into $b = 6$ batches.

Table 2.7 lists the experimental results obtained from sampling observations from three different i.i.d. processes. The μ row lists the true mean. The avg samp row lists the average sample size. The avg and stdv hw rows list, respectively, the average half-width and standard deviation of the half-width obtained by the procedure. The coverage row lists the percentage of the c.i.'s that cover the true mean value. The c.i. coverages are around the specified 90% confidence level. The procedure correctly detects that the underlying sequences are independent and the size of allocated samples is around the theoretical value, which is approximated 200 ($=1.\bar{1} \times 180$).

Table 2.8 lists the experimental results from more-complicated stochastic processes. The steady-state distributions of the MA1 and AR1 processes are normal and the c.i. coverages are around the specified 90% confidence level. For the M/M/1 queuing process, samples are not only highly correlated but also far from normal; the steady-state distribution of the M/M/1 queuing process is not only asymmetric but also discontinuous at $x = 0$. Hence, the lag l of the systematic samples of the M/M/1 queuing process to appear independent distributed is significantly larger than that of the MA1 and AR1 processes. The coverage of the M/M/1 queuing process is slightly lower than that of the MA1 and AR1 processes.

2.9 Summary

We have presented an algorithm for determining the simulation run length, and a strategy for building a c.i. for the mean μ of a steady-state simulation response. The procedure estimates the required sample size based entirely on data and does not require user intervention and can easily be incorporated into any simulation procedures. The experimental evaluation reveals that the procedure determines batch sizes that are sufficiently large for achieving adequate c.i. coverage. For independent sequences, the QI procedures will allocate a small sample size (a little more than 200) and deliver a valid c.i. with the default precision. As the autocorrelation of the sequences increases, the procedures will increase the sample size to compensate for the lack of independence. Using a straightforward test of independence to determine the simulation run length making the procedure easy to understand and simple to implement. Furthermore, the procedure needs only to process each observation once and does not require storing the entire output sequence. As such, the QI procedure is an online algorithm, it can process its input piece-by-piece in a serial fashion without having the entire input sequence available from the start. Online algorithm is needed in situations where decisions must be made and resources allocated without knowledge of the future.

Chapter 3

Generating Independent and Identically Distributed Batch Means

Many simulation procedures are derived based on the assumption that data are i.i.d. normal; for example, ranking and selection procedures and multiple-comparison procedures. Batch means are the method of choice to manufacture data that are approximately i.i.d. normal. In this chapter, we discuss a procedure to manufacture batch means that appear to be i.i.d. normal, as determined by the von Neumman test of independence and chi-square test of normality. The procedure performs statistical analysis on sample sequences collected on strictly stationary stochastic processes. It determines the batch sizes based entirely on data and does not require any user intervention. The only required condition is that the autocorrelations of the stochastic process output sequence die off as the lag between observations increases, in the sense of ϕ-mixing.

While some methods attempt to estimate the SSVC for construction of a c.i., non-overlapping batch means (NOBM) in its classical setting (i.e., when the number of batches is fixed) does not. The presented QIN method of [Chen and Kelton (2007)] is a classical NOBM method and it does not estimate the SSVC.

3.1 Discussion of Batch-Means Method

There are a number of batch-size-determination procedures that aim to manufacture independent batch means. However, these methods have focused on selecting a batch size large enough to achieve near independence of the batch means and ignored the question of normality based on the assumption that if the batch size is large enough for the batch means to be approximately independent, then the batch size is large enough for the batch means to be approximately normally distributed. The procedure of

[Law and Carson (1979)] starts with 400 batches of size 2 and doubles sample sizes every two iterations until an estimate for lag-1 correlation among 400 batch means becomes smaller than 0.4 and larger than the estimated lag-1 correlation among 200 batch means with twice the batch size. A drawback of the method is that it does not address the issue of the normality of the batch means. Another widely studied batch-means procedures are the set of methods LBatch and ABatch, see [Fishman (2001)]. However, these procedures require users enter the simulation run length at the beginning of the execution. Hence, even though these procedures dynamically determine the optimal batch sizes (so that batch means converge faster to normality), they are fixed-sample-size procedures. [Schmeiser (1982)] reviews several batch-means procedures and concludes that selecting between 10 and 30 batches should suffice for most simulation experiments.

3.2 Generating Independent and Normally Distributed Batch Means

In this section, we discuss the strategy of estimating the required batch size so that batch means appear to be independent and normally distributed as well as a procedure to construct confidence interval of the mean. [Chen and Kelton (2007)] denote the method quasi-independent-and-normal (QIN) procedure.

3.2.1 *Validation of Normality*

Batch means that appear to be independent are not necessarily normally distributed and vice versa. However, our experimental results indicate that c.i.'s constructed with the assumption that samples are i.i.d. normal generally have coverages close to the nominal value when samples are independent but not normal. Therefore, it is not as critical to ensure batch means are normally distributed as to ensure batch means are independent in terms of c.i. coverage. Nevertheless, ensure batch means are normal can improve the c.i. coverage and shorten the half-width. Furthermore, some procedures require data are i.i.d. normal.

Let $z_{1-\alpha}$ be the $1-\alpha$ quantile of the standard normal distribution. To determine whether batch means appear to be normally distributed, we check the proportions of the value of batch means in each interval bounded by (-∞, $\hat{\mu} - z_{0.8333}S$, $\hat{\mu} - z_{0.6667}S$, $\hat{\mu}$, $\hat{\mu} + z_{0.6667}S$, $\hat{\mu} + z_{0.8333}S$, ∞), where

$\hat{\mu}$ and S are, respectively, the grand sample mean of these n observations and standard error of these b batch means.

These intervals are strategically chosen so that the proportions of batch means in each interval are approximately equal. Under the null hypothesis that the batch means are normally distributed, the proportions of batch means in each interval are approximately (0.1667, 0.1666, 0.1667, 0.1667, 0.1666, 0.1667). We apply the chi-square test of normality, see [Law (2014)], to these batch means. We use a 0.9 confidence level for the chi-square test.

The powers of the independence and the normality tests increase as the number of batches used to perform the test increases. Based on previous experimental results we recommend the sample size used for this normality test and the von Neumann test of independence be at least 180 for the intend that we are using these tests. See Section 3.3 for some experimental results of the chi-square test of normality and the von Neumann test of independence. There are other more sophisticated normality tests, for example, the Shapiro-Wilk test of normality [Bratley et al. (1987)]. We chose chi-square test because it is easy to apply and serves the purpose well, however, other normality tests can be used in place of the chi-square test of normality in the method if users desire.

3.2.2 *A Source Code of Normality test*

This subroutine implements the normality test.

```
/* normality test */
int normality(double* BatchN, int bnum,
              double muhat, double stdev)
{
/* proportion of each proportion */
static double p[6] = {0.1667,0.1666,0.1667,
                      0.1667,0.1666,0.1667};
int i;
int n[6] = {0, 0, 0, 0, 0, 0};
int df = 5;

   for (i = 0; i < bnum; ++i) {
     if (BatchN[i] < muhat - quantile_normal(0.8333)*stdev)
         n[0] = n[0] + 1;
```

```
      else
      if (BatchN[i] < muhat - quantile_normal(0.6667)*stdev)
           n[1] = n[1] + 1;
      else
      if (BatchN[i] < muhat)
           n[2] = n[2] + 1;
      else
      if (BatchN[i] < muhat + quantile_normal(0.8333)stdev)
           n[3] = n[3] + 1;
      else
      if (BatchN[i] < muhat + quantile_normal(0.6667)*stdev)
           n[4] = n[4] + 1;
      else
           n[5] = n[5] + 1;
    }; /* for */

    if ( chisquare_test(n, p, df) ) {
       return (1);
    } else {
       return (0);
    };
}
```

3.2.3 *Batch Means Variance Estimator*

[Fishman (2001)] classifies the difference of $S^2 - \sigma^2$ into three categories: 1) error due to finite sample size n; 2) error due to ignoring correlation between batches; 3) error due to random sampling. He collectively refers to the errors in the first two categories as *systematic variance error*. Recall that in NOBM $n = bm$, where b is the number of batches and m is the batches size. He then points out that under relatively weak conditions, the systematic variance error behaves as $O(1/m)$, and the standard error of error due to random sampling behaves as $O(1/\sqrt{b})$. Here $O(a_n)$ denotes a quantity that converges to zero at least as fast as does the sequence $\{a_n\}$, as $n \to \infty$. Hence, using a fixed number of batches the batch size $m \propto n$ would diminish the systematic variance error most rapidly. On the other hand, if we want to reduce the error due to random sampling, we should increase the number of batches.

The basic idea of QIN is that Eq. (2.3) has become asymptotically valid when the batch means $\bar{X}_j$, for $j = 1, 2, \ldots, b$, appear to be independent and normally distributed, as determined by the von Neumann test and chi-square test. That is, the degree of systematic variance error has diminished to a level that can be neglected. Hence, we don't need to increase the batch size any further since the goal is to manufacture as many as i.i.d. normal batch means as possible with a given sample size.

If $\{X_i\}$ is an i.i.d. $N(\mu, \sigma^2)$ sequence, then the $1-\alpha$ half-width of sample mean constructed with n observations is

$$w = z_{1-\alpha/2}\frac{\sigma}{\sqrt{n}}.$$

If the sequence is divided into b non-overlapping batch means with batch size m, i.e., $n = bm$, then the variance of batch means is σ^2/m and the $1 - \alpha$ half-width constructed with b batch means is

$$w = z_{1-\alpha/2}\frac{\sigma/\sqrt{m}}{\sqrt{b}} = z_{1-\alpha/2}\frac{\sigma}{\sqrt{n}}.$$

That is, for i.i.d. normal sequences the batch size has no impact on the c.i. half-width w when the sample size is fixed. However, this property generally does not hold when the variance is unknown, and the c.i. half-width needs to be estimated by

$$w = t_{1-\alpha/2,b-1}\frac{S}{\sqrt{b}}. \tag{3.1}$$

[Chien et al. (1997)] show that under certain assumptions:

$$\text{Var}[S^2] \to \frac{2\sigma^4(b+1)}{(b-1)^2} + O(1/b^2) \text{ as } b \to \infty. \tag{3.2}$$

[Fishman (2001)] shows that if S^2 converges to σ^2 in mean square, then for fixed α the ratio of expected widths as a function of the number of batches b would asymptotically be

$$\frac{t_{b-1,1-\alpha/2}}{z_{1-\alpha/2}}[1 - \frac{b+1}{2(b-1)^2}] \geq 1, \forall b \geq 2.$$

Once the QIN algorithm has determined that the sample size is large enough for the asymptotic approximation to become valid, we then compute the mean and variance based on these batch means. The mean and the c.i. half-width are estimated, respectively, by $\bar{X}$ and Eq. (3.1). The final step in the procedure is to determine whether the c.i. meets the user's half-width requirement, a maximum absolute half-width ϵ or a maximum relative fraction γ of the magnitude of the final point mean estimator $\bar{X}$. If the relevant requirement $w \leq \epsilon$, or $w \leq \gamma|\bar{X}(n)|$ for the precision of the c.i. is satisfied, then the procedure terminates, and we return the point estimator $\bar{X}(n)$ and the c.i. with half-width w. If the precision requirement is not satisfied, then the procedure will increase the number of batches according to Eq. (2.4).

3.2.4 *The Implementation*

The procedure progressively increases the batch size until the batch means appear to be independent, as determined by the von Neumann test of independence, and appear to be normally distributed, as determined by the chi-square test. We divide the entire output sequence into $b = 180$ batches. To reduce the storage requirement, we allocate a buffer with size $3b$ to keep sample means, i.e., the average of l observations. Batch means are then computed from these sample means in the buffer. Initially each observation is treated as a sample mean; as the procedure proceeds the number of observations used to compute these sample means will be doubled every two iterations.

The following shows certain properties at each iteration:

Iteration	0	1_A	1_B	2_A	2_B	...	r_A	r_B
Total Observations	b	$2b$	$3b$	$4b$	$6b$	...	$2^k b$	$2^{k-1}3b$
No. of Sample Means	b	$2b$	$3b$	$2b$	$3b$	...	$2b$	$3b$
No. of Observations	1	1	1	2	2	...	2^{k-1}	2^{k-1}
Batch Size	1	2	3	4	6	...	2^k	$2^{k-1}3$

The *Iteration* row shows the index of the iteration. The *Total Observations* row shows the total number of observations at a certain iteration. The *No. of Sample Means* row shows the total number of sample means in the buffer at a certain iteration. The *No. of Observations* row shows the number of observations used to obtain the sample means in the buffer, i.e., the value of l. The *Batch Size* row shows the number of observations used to compute the batch means, i.e., the batch size. Note that the sample mean here could be the value of one observation.

There maybe b, $2b$, or $3b$ sample means in the buffer. We aggregate the available sample means into b batch means by averaging the adjacent sample means. For example, at the end of the 2_A^{th} iteration, the batch size is 4. We compute batch means by averaging 2 consecutive samples means that are the average of 2 consecutive samples. The procedure will progressively increase l (the number of observations used to obtain sample means in the buffer) and consequently the batch size until these b batch means appear to be independent and normally distributed.

The quasi-independent-and-normal algorithm:

(1) Remark: The size of the buffer used to store the sample means is $B_s = 3b$, l is the number of observations used to compute sample means in the buffer, r is the index of iterations. Each iteration r contains two sub-iterations r_A and r_B.
(2) Set $b = 180$, $l = 1$, and $r = 0$. Generate b observations as the initial samples.
(3) If this is the initial iteration, set the value of batch means to the initial b sample means in the buffer. If this is a r_A iteration, set the value of batch means to the average of two consecutive sample means in the buffer. If this is a r_B iteration, set the value of batch means to the average of three consecutive sample means in the buffer.
(4) Carry out tests to determine whether these b batch means appear to be independent and normally distributed. If the batch means appear to be independent and normal, go to step 10.
(5) If the current iteration is the initial or r_B iterations, set $r = r + 1$ and start a r_A iteration. If the current iteration is r_A iterations, start a r_B iteration.
(6) If this is the 1_A^{th} or 1_B^{th} iteration, generate b observations, store those values in the buffer after the ones already there and go to step 3.
(7) If this is a r_A iteration ($r > 1$), then re-calculate the sample means in the buffer by taking the average of two consecutive sample means, reindex the rest of $3b/2$ sample means in the first half of the buffer. Generate another $b/2$ sample means; each sample mean is the average of consecutive $l = 2^{r-1}$ observations and store those values in the later portion of the buffer.
(8) If this is a r_B iteration ($r > 1$), generate another b sample means; each sample mean is the average of consecutive $l = 2^{r-1}$ observations and store those values in the later portion of the buffer.
(9) Go to step 3.
(10) Compute the variance estimator according to Eq. (2.3) and confidence interval half-width according to Eq. (3.1).
(11) Let ϵ be the desired absolute half-width and $\gamma|\bar{X}|$ be the desired relative half-width. If the half-width of the c.i. is greater than ϵ or $\gamma|\bar{X}|$, compute b', the required number of batches by Eq. (2.4), generate

$b' - b$ additional batches, set $b = b'$, and go to step 10; otherwise the procedure returns the c.i. estimator and terminates.

The QIN procedure starts with an initial sample size of 180 and doubles the sample sizes every two iterations. The procedure progressively increase the batch size so that these batch means appear to be independent and normally distributed. Hence, we can use these batch means to construct a classical confidence interval without any adjustment. The procedure needs only process each observation once and does not require storing the entire output sequence. That is, the QIN procedure is an online algorithm.

The main purpose of the procedure is to manufacture batch means that appear to be independent and normally distributed. If users are more interested in using these batch means to construct c.i.'s, then smaller number of batches with larger batch size can be used. In general, with a fixed-length sequence the c.i. half-width becomes smaller as the number of batches increases. For example, it is likely that the constructed c.i. will have great coverage with large c.i. half-width if we divide the entire sequence into two batches. For details on batch-size effects in the analysis of simulation output once the sample size is fixed, see [Schmeiser (1982)].

3.2.5 *Discussions of Batch-Means Procedures*

In this section, we discuss the rationale of the QIN procedure and point out the difference and similarity between QIN and the procedure of [Law and Carson (1979)], denoted LC and the set of methods LBatch and ABatch [Fishman (2001)].

The sample size incremental strategy of QIN is very similar to that of LC; LC doubles the sample sizes every two iterations, by setting $n_0 = 600$, $n_1 = 800$, and $n_i = 2n_{i-2}$. However, the stopping criteria are different. Once the stopping criteria are satisfied, LC constructs c.i. by dividing the entire sequence into 40 batches. LC aggregates every 10 batch means that appear to be independent as the final batch means and does not explicitly check whether these batch means appear to be normally distributed. Furthermore, if the obtained c.i. half-width is wider than desired, LC will keep the number of batches at 40 and increases only the batch size. Note that in the implementation we only need to process each observation once.

The set of methods LBatch and ABatch [Fishman (2001)] incorporates two different sample size incremental strategies: FNB (Fixed Number of Batches) and SQRT (the number of batches and batch size are increased by $\sqrt{2}$ at each iteration). Both FNB and SQRT double the sample size at each iteration. However, LBatch and ABatch require users enter the sample size n for a simulation run. Let b be the initial number of batches. If n equals n', the minimal sample size for the batch means with batch size n'/b to appear to be independent, then the batch sizes determined by LBatch and ABatch are the same since these two procedures use FNB rule exclusively and invoke SQRT rule only after batch means appear to be independent. [Fishman (2001)] points out that batch-means methods that based entirely on the FNB rule, for example LC, do obtain asymptotically valid variance estimator S^2, however, they do not converge in mean square to σ^2 and are not statistically efficient.

QIN also uses FNB rule, however, it doubles the sample size every two iterations instead of every iteration. Moreover, QIN continues to use FNB rule to increase batch size until batch means appear to be normally distributed. However, after QIN estimates a sufficient large batch size such that batch means appear to be independent and normally distributed, it does not increase the batch size any further. For example, if the obtained c.i. half-width is wider than desired, QIN will only increase the number of batches with the estimated batch size, which will reduce the variation of the variance estimator more rapidly, see Eq. (3.2). On the other hand, if $n > n'$ LBatch and ABatch will increase both the number of batches and batch sizes. Note that LBatch and ABatch do not explicitly check whether batch means appear to be normally distributed and SQRT rule can ensure a faster convergent to normality with a given n.

3.3 Empirical Experiments

In this section, we present some empirical results from simulation experiments using the proposed procedure. We use 180 batch means for the von Neumann test of independence and chi-square test of normality. We test the procedure with six stochastic processes:

- Observations are i.i.d. uniform between 0 and 1, denoted U(0,1).
- Observations are i.i.d. normal with mean 0 and variance 1, denoted $N(0,1)$.
- Observations are i.i.d. exponential with mean 1, denoted expon(1).
- Steady-state of the *first-order moving average* process, generated by

the recurrence

$$X_i = \mu + \epsilon_i + \theta\epsilon_{i-1} \text{ for } i = 1, 2, \ldots,$$

where ϵ_i is i.i.d. $N(0, 1)$ and $0 < \theta < 1$. μ is set to 0 in the experiments. This process is denoted MA1(θ).

- Steady-state of the *first-order autoregressive* process, generated by the recurrence

$$X_i = \mu + \varphi(X_{i-1} - \mu) + \epsilon_i \text{ for } i = 1, 2, \ldots,$$

where ϵ_i is i.i.d. $N(0, 1)$, and $0 < \varphi < 1$. μ is set to 0 and X_0 is set to a random variate drawn from the steady-state distribution in the experiments. This process is denoted AR1(φ).
- Steady-state of the M/M/1 delay-in-queue process with the arrival rate (λ) and the service rate ($\nu = 1$). This process is denoted MM1(ρ), where $\rho = \lambda/\nu$ is the traffic intensity.

We carry out experiments to evaluate the performance of using the von Neumann test to determine whether these batch means appear to be independent, and using the chi-square test to determine whether these batch means appear to be normally distributed. We check the batch size at which these batch means appear to be independent and normally distributed, and check the interdependence between the batch size at which these batch means appear to be independent and the strength of the autocorrelation of the output sequence. The confidence level of the von Neumann test and the chi-square test is set to 90% in the experiments. Note that a lower confidence level of these tests will increase the batch size and the simulation run length. We evaluate the performance of using these approximately i.i.d. normal batch means to estimate the variance of sample means.

3.3.1 *Experiment 1: Independence and Normality Tests*

In this experiment, we used the von Neumann test to determine whether a sequence appears to be independent and the chi-square test to determine whether a sequence appears to be normal. It can be viewed as batch size $m = 1$ and the number of batches $b = 180$. Tables 3.1 and 3.2 lists the experimental results. Each design point is based on 10,000 independent simulation runs. The $P(100(1 - \alpha_{ind})\%)$ columns under Independence list the observed percentage of correct decision when the nominal probability of the von Neumann test is set to $1 - \alpha_{ind}$, i.e., the percentage of these 10,000 runs passing the von Neumann test of independence for the U(0,1), $N(0, 1)$,

Table 3.1 Test of independence with 180 batch means

Test	Independence		
Process	$P(90.00\%)$	$P(95.00\%)$	$P(99.00\%)$
U(0,1)	89.91%	94.96%	99.00%
$N(0,1)$	90.29%	95.19%	98.94%
expon(1)	90.23%	94.75%	98.54%
MA1(0.15)	74.97%	61.45%	34.77%
MA1(0.25)	97.51%	93.97%	80.28%
MA1(0.35)	99.95%	99.77%	98.08%
MA1(0.50)	100.00%	100.00%	99.98%
MA1(0.75)	100.00%	100.00%	100.00%
MA1(0.90)	100.00%	100.00%	100.00%
AR1(0.15)	75.72%	62.58%	36.60%
AR1(0.25)	97.93%	95.13%	84.13%
AR1(0.35)	99.97%	99.90%	99.15%
AR1(0.50)	100.00%	100.00%	99.99%
AR1(0.75)	100.00%	100.00%	100.00%
AR1(0.90)	100.00%	100.00%	100.00%
MM1(0.15)	77.29%	72.47%	62.13%
MM1(0.25)	96.27%	94.83%	90.65%
MM1(0.35)	99.72%	99.54%	98.87%
MM1(0.50)	100.00%	99.99%	99.98%
MM1(0.75)	100.00%	100.00%	100.00%
MM1(0.90)	100.00%	100.00%	100.00%

and expon(1) design points and failing the von Neumann test for all other design points. The $P(100(1-\alpha_{nor})\%)$ columns under Normality list the observed percentage of correct decision when the nominal probability of the chi-square test of normality is set to $1-\alpha_{nor}$, i.e., the percentage of these 10,000 runs passing the chi-square test for the $N(0,1)$, MA1(θ), and AR1(φ) design points and failing the chi-square test for all other design points. We set α_{ind} and α_{nor} to 0.1, 0.05, and 0.01.

A Type I error of a hypothesis test is the event that we reject the null hypothesis when it is true, i.e., independent sequences fail the von Neumann test. The von Neumann test performs very well in terms of Type I error, it is around the specified α_{ind} level. A Type II error is the event that we accept the null hypothesis when it is false, i.e., correlated sequences pass the test of independence. When the sequence is only slightly correlated, Type II error is high, for example, the MA1(0.15), AR1(0.15), and MM1(0.15) processes. Note that we can use a larger number of batches to increase the power of test and reduce Type II error. If we mistakenly treat slightly correlated sequences as being i.i.d., the performance measurements should still be fairly accurate, thus, the low probability of correct decision of those slightly correlated sequences should not pose a problem. On the

Table 3.2 Test of normality with 180 batch means

Test	Normality		
Process	$P(90.00\%)$	$P(95.00\%)$	$P(99.00\%)$
U(0,1)	80.41%	65.46%	31.86%
$N(0,1)$	96.64%	98.30%	99.72%
expon(1)	99.93%	99.82%	99.42%
MA1(0.15)	96.28%	98.37%	99.72%
MA1(0.25)	96.11%	98.25%	99.69%
MA1(0.35)	96.36%	98.40%	99.68%
MA1(0.50)	96.56%	98.52%	99.81%
MA1(0.75)	96.56%	98.58%	99.78%
MA1(0.90)	96.34%	98.43%	99.73%
AR1(0.15)	96.65%	98.40%	99.70%
AR1(0.25)	96.26%	98.23%	99.74%
AR1(0.35)	96.44%	98.46%	99.83%
AR1(0.50)	96.35%	98.32%	99.70%
AR1(0.75)	93.54%	96.70%	99.10%
AR1(0.90)	78.31%	86.01%	94.20%
MM1(0.15)	100.00%	100.00%	100.00%
MM1(0.25)	100.00%	100.00%	100.00%
MM1(0.35)	100.00%	100.00%	100.00%
MM1(0.50)	100.00%	100.00%	100.00%
MM1(0.75)	95.90%	93.73%	87.21%
MM1(0.90)	83.08%	77.27%	65.02%

other hand, when the output sequences are mild to highly correlated, the von Neumann test detects the dependence almost all the time. The results of the normality test indicate that the autocorrelations among the samples has very little impact on the chi-square normality test when the autocorrelations are small. However, as the autocorrelations become stronger the chi-square normality test starts to break down, for example, the AR1(0.9) and MM1(0.9) processes. Even though the von Neumann test of independence does not guarantee independence, the sequence passes the test will only be slightly correlated if it is not independent. Hence, the residual correlation (if there is any) left in the batch means will have little impact on the chi-square normality test in the QIN procedure. Since $\alpha_{ind} = 0.1$ and $\alpha_{nor} = 0.1$ have a better performance in balancing Type I and II errors, we chose $\alpha_{ind} = 0.1$ and $\alpha_{nor} = 0.1$ for these tests in the procedure.

3.3.2 *Experiment 2: Batch Sizes Determination*

In this section, we check the batch size at which the batch means appear to be normally distributed and the batch size at which the batch means appear to be independent with respect to the strength of the autocorrelation of

Table 3.3 Average batch size m at which these 180 batch means appear to be normally distributed (pass the chi-square normality test)

Process	U(0, 1)		N(0, 1)		expon(1)	
	$\bar{m}$	$stdv(m)$	$\bar{m}$	$stdv(m)$	$\bar{m}$	$stdv(m)$
	1.85	0.46	1.04	0.19	3.96	1.30

the output sequence. Each design point is based on 10,000 independent simulation runs.

Table 3.3 lists the experimental results. The $\bar{m}$ column lists the average batch size at which the batch means appear to be normally distributed. The $stdv(m)$ column lists the standard deviation of the batch size m. For example, in average with batch size of 1.85 U(0, 1) observations appear to be normally distributed. The procedure correctly detects the normality of the observations sampled from the $N(0, 1)$ distribution; hence, the average batch size m is 1.04 close to the theoretical value of 1. This is consistent with the result in experiment 1 that the test of normality encounters Type I error approximately 4% of the time with $\alpha_{nor} = 0.1$, consequently, the average batch size is approximately $1 \times 0.96 + 2 \times 0.04$.

Table 3.4 lists the experimental results of the batch size at which the batch means appear to be independent. The θ, φ, ρ column lists the coefficient values of the corresponding stochastic process. The MA1(θ) column lists the results of the underlying moving average output sequences. In general, the average batch size at which the batch means appear to be independent increases as the autocorrelation increases. The MA1(θ) processes are only slightly correlated even with θ as large as 0.9. Therefore, in average batch means with batch size of 4 of the MA1(0.90) output sequences appear to be independent. On the other hand, the MM1(0.90) output sequences are highly correlated, the average batch size at which the batch means appear to be independent is as large as 321. The results from this experiment indicate a strong correlation between the average batch size m at which the batch means appear to be independent and the strength of the autocorrelation.

We also performed the experiment using different number of batches for the underlying independence and normality tests. The average batch size at which the batch means appear to be independent and normal generally increases as the number of batches used for those tests increases since power of test increases as sample size increases. We choose to use sample size of 180 for these tests because based on previous experimental tests using this sample size meets the precision requirement and does not cause the

Table 3.4 Average batch size m at which these 180 batch means appear to be independent (pass von Neumann test of independence)

Process	MA1(θ)		AR1(φ)		MM1(ρ)	
θ,φ,ρ	$\bar{m}$	$stdv(m)$	$\bar{m}$	$stdv(m)$	$\bar{m}$	$stdv(m)$
0.15	2.07	0.86	2.22	0.99	2.78	1.48
0.25	2.64	0.89	3.43	1.21	5.16	2.09
0.35	2.93	0.96	4.65	1.39	8.20	2.57
0.50	3.22	1.04	6.91	1.73	15.54	4.03
0.75	3.44	1.09	14.77	2.87	60.74	12.46
0.90	3.48	1.10	34.68	5.46	320.68	55.58

simulation to run longer than necessary.

3.3.3 *Experiment 3: Coverages of Confidence Interval*

In this experiment, we use the QIN procedure to determine batch sizes so that batch means appear to be independent and normally distributed. Since batch means are often used to estimate the variance of sample means, we evaluate the accuracy of the variance estimated by the procedure. In these experiments, no relative precision or absolute precision were specified, so the half-width of the c.i. is the result of the default precision. Furthermore, based on the common rule of thumb that the optimal number of batch means is around 30 to 40 when using batch means to estimate variance and the sample size is fixed, in this experiment we list the results of c.i. coverage when the entire sequences are divided into 180 batches as well as 30 batches.

Table 3.5 lists the experimental results of sampling independently from three different distributions. The μ row lists the true mean. The $\bar{X}$ row lists the grand sample mean. The avg and stdv rp rows list, respectively, the average relative precision and standard deviation of the relative precision of the point estimator. Here, the relative precision is defined as $\gamma = |\bar{X} - \mu|/\bar{X}$, where $|x|$ is the absolute value of x. The avg and stdv samp rows list, respectively, the average sample size and standard deviation of the sample size. The avg and stdv bsize rows list, respectively, the average batch size and standard deviation of the batch size obtained by the procedure. The avg and stdv hw rows list, respectively, the average half-width and standard deviation of the half-width obtained by the procedure. The coverage row lists the percentage of the c.i.'s that cover the true mean value. The average batch size is 1.19 when sampling from i.i.d. normal distribution, which is close to the theoretical value of 1 and reflects the fact that the underlying observations are independent and normally distributed.

Table 3.5 Coverage of 90% confidence intervals of independent samples

Process	$U(0,1)$	$N(0,1)$	$expon(1)$
μ	0.50	0.00	1.00
$\bar{X}$	0.499828	-0.000299	1.000176
avg rp	0.025552	1.0	0.028747
stdv rp	0.020226	0.0	0.023740
avg samp	365	214	973
stdv samp	166	111	800
batches		180	
avg bsize	2.03	1.19	5.41
stdv bsize	0.92	0.62	4.45
avg hw	0.026186	0.117864	0.058698
stdv hw	0.004453	0.014815	0.013726
coverage	90.09%	90.57%	89.85%
batches		30	
avg bsize	12.17	7.12	32.44
stdv bsize	5.54	3.70	26.68
avg hw	0.026469	0.119077	0.059236
stdv hw	0.005379	0.020267	0.015295
coverage	89.97%	90.25%	89.75%

On the other hand, the batch means do not appear to be normal until the average batch size is 2.03 when sampling from U(0,1) distribution. Because exponential distribution is asymmetric, the batch means do not passed the normality test until the average batch size is 5.41. The c.i. coverages are around the specified 90% confidence level. For independent observations, it seems that the batch sizes determined by the procedure are large enough for these batch means to be normally distributed so that there is no significant difference in coverages when batch sizes are increased 6 times. This is consistent with the discussion in Section 3.2.3.

Table 3.6 lists the experimental results of the chosen stochastic processes. For these three tested processes, the c.i. coverages are around the specified 90% confidence level. Since the steady-state distribution of the MA1 and AR1 processes are normal, we believe that some of the batch means that passed the test of independence may be slightly correlated. For the M/M/1 queuing process, samples are not only highly correlated but also far from normal; the steady-state distribution of the M/M/1 queuing process is not only asymmetric but also discontinuous at $x = 0$. Hence, the batch size for the M/M/1 queuing process to appear independent and normally distributed is significantly larger than that of the MA1 and AR1 processes. Consequently, the QIN procedure will allocate a large sample

Table 3.6 Coverage of 90% confidence intervals of correlated samples

Process	MA1(0.9)	AR1(0.9)	MM1(0.9)
μ	2.00	2.00	9.00
$\bar{X}$	1.998995	1.998049	8.989608
avg rp	0.028616	0.036842	0.012098
stdv rp	0.023504	0.031344	0.010345
avg samp	830	14560	3296931
stdv samp	980	12481	2992981
batches		180	
avg bsize	4.61	80.89	18316.29
stdv bsize	5.45	69.34	16627.67
avg hw	0.113788	0.140637	0.201692
stdv hw	0.021196	0.026812	0.064079
coverage	88.4%	88.1%	87.0%
batches		30	
avg bsize	27.66	485.33	109897.73
stdv bsize	32.68	416.05	99766.04
avg hw	0.119555	0.147523	0.203708
stdv hw	0.026072	0.033256	0.067350
coverage	89.3%	89.2%	86.7%

size and deliver a tight c.i. when underlying processes are asymmetric and discontinuous. In addition to some correlated batch means pass the test of independence, some non-normal batch means pass the test of normality, thus, the coverage of the M/M/1 queuing process is slightly lower than that of the MA1 and AR1 processes.

3.4 Summary

We have presented an algorithm for estimating the required batch size so that batch means appear to be i.i.d. normal and a strategy for building a c.i. on a steady-state simulation response. The QIN algorithm works well in determining the required batch size for the asymptotic approximation of batch means to become valid. The procedure estimates the required sample size based entirely on data and does not require any user intervention. Moreover, the QIN procedure needs to process each observation only once and does not require storing the entire output sequence so the storage requirement is minimal. The QIN procedure can be used as a pre-processor of simulation procedures that require i.i.d. normal data. The experimental evaluation reveals that QIN determines batch sizes that are sufficiently large for achieving approximately i.i.d. normal batch means and for achieving adequate c.i. coverage. The main advantage of the approach is that by using a straightforward test of independence and test of normality to determine the valid batch size so that batch means appear to be independent and normally distributed, we can apply classical statistical techniques directly and do not require more advanced statistical theory, thus making it easy to understand and simple to implement.

Chapter 4

Distributions of Order Statistics

Order statistics are among the most fundamental tools in non-parametric statistics and inference and have been widely studied and applied to many real-world issues. It is often of interest to estimate the reliability of the component/system from the observed lifetime data. In many applications we want to estimate the probability that a future observation will exceed a given high level during some specified epoch. For example, a machine may break down if a certain temperature is reached.

Let X_i, $i = 1, 2, \ldots, n$, denote a sequence of mutually independent random samples from a common distribution of the continuous type having a probability density function (pdf) f and a cumulative distribution function (cdf) F. Let $X_{[u]}$ be the uth smallest of these X_i such that $X_{[1]} \leq X_{[2]} \leq \cdots \leq X_{[n]}$. Then $X_{[u]}$, $u = 1, 2, \ldots, n$, is called the uth order statistic of the random sample X_i, $i = 1, 2, \ldots, n$. Note that though the samples $X_1, X_2, \cdots, X_n$ are i.i.d., the order statistics $X_{[1]}, X_{[2]}, \cdots, X_{[n]}$ are not independent because of the order restriction. The difference $R = X_{[n]} - X_{[1]}$ is called the sample range. It is a measure of the dispersion in the sample and should reflect the dispersion in the population. For more information on order statistics see [David and Nagaraja (2003)].

Suppose $U_1, U_2 \sim U(0, 1)$, where "$\sim$" denotes "is distributed as" and "$U(a, b)$" denotes a uniform distribution with range $[a, b]$. We are interested in the distribution of $U_{[2]} = \max(U_1, U_2)$, which can be viewed as the second order statistic. The cdf of $U_{[2]}$ is

$$\begin{aligned} \Pr(\max(U_1, U_2) \leq x) &= \Pr(U_1 \leq x, U_2 \leq x) \\ &= \Pr(U_1 \leq x)\Pr(U_2 \leq x) \\ &= x^2. \end{aligned}$$

Furthermore, the cdf of $U_{[1]} = \min(U_1, U_2)$, which can be viewed as the first order statistic, is

$$\begin{aligned}\Pr(\min(U_1, U_2) \le x) &= 1 - \Pr(\min(U_1, U_2) > x)\\ &= 1 - \Pr(U_1 > x)\Pr(U_2 > x)\\ &= 1 - (1 - x)^2.\end{aligned}$$

Consider a device contains a series of two identical components and the lifetime of the component $L_c \sim U(0, 1)$. If any of these two components fail, the device fails, such as chains. Hence, the lifetime of the device $L_d \sim \min(L_{c_1}, L_{c_2})$. On the other hand, if these two components are configured in parallel and the device fails only when both components fail, such as a strand contains a bundle of threads. Then the lifetime of the device $L_d \sim \max(L_{c_1}, L_{c_2})$. These techniques can be used in reliability analysis to estimate the reliability of the component/system from the observed lifetime data. When there are more than two samples, the distributions of order statistics can be obtained via the same technique.

[Hogg et al. (2012)] show that the distribution of the uth order statistic of n samples of X (i.e., $X_{[u]}$) is

$$g_{u:n}(x) = \beta(F(x); u, n - u + 1)f(x),$$

where

$$\beta(x; a, b) = \frac{1}{\beta(a, b)} x^{a-1}(1 - x)^{b-1} \text{ and } \beta(a, b) = \frac{\Gamma(a)\Gamma(b)}{\Gamma(a + b)}$$

is the beta function with shape parameters a and b. Note that $\Gamma(a) = (a - 1)!$ for any positive integer a. Furthermore, the cdf

$$G_{u:n}(x) = \int_{-\infty}^{x} g_{u:n}(y)dy.$$

The cdf $G_{u:n}(x)$ can also be expressed as

$$\begin{aligned}\Pr(X_{[u]} \le x) &= \Pr(\text{at least } u \text{ of the } X_i \text{ are not greater than } x)\\ &= \sum_{i=u}^{n} \binom{n}{i} F^i(x)(1 - F(x))^{n-i}.\end{aligned}$$

The last equality holds because each term in the summation corresponds to the probability that exactly i of the $X_1, X_2, \cdots, X_n$ are not greater than x.

In particular, the first order statistic, or the sample minimum, $X_{[1]}$ has the pdf

$$g_{1:n}(x) = n[1 - F(x)]^{n-1} f(x)$$

and the cdf

$$G_{1:n}(x) = 1 - [1 - F(x)]^n.$$

The nth order statistic, or the sample maximum, $X_{[n]}$ has the pdf

$$g_{n:n}(x) = n[F(x)]^{n-1} f(x)$$

and the cdf

$$G_{n:n}(x) = [F(x)]^n.$$

In the case that f is the uniform [0,1] distribution,

$$g_{u:n}(x) = \beta(x; u, n - u + 1).$$

That is, let $x = U \sim U(0,1)$, then the random variate $y = \beta(U; u, n-u+1)$, is the uth order statistic of n random variables with uniform $[0,1]$ as the parent distribution. From this we can deduce that

$$E[X_{[u]}] = \frac{u}{n+1}, \tag{4.1}$$

and

$$\text{Var}[X_{[u]}] = \frac{u(n-u+1)}{(n+1)^2(n+2)}.$$

This result can be used to generate the uth order statistic of any variates with a strictly increasing cdf. A straightforward way of simulating order statistics is to generate a pseudorandom sample from the distribution $F(X)$ and then sort the sample in ascending order. This general method requires sorting and can be avoided by using the probability integral transform. The probability integral transform or transformation states that data values that are modelled as random variables from any given continuous distribution can be converted to random variables having a uniform distribution. Based on the fact that $y = F(x) \sim U(0,1)$, we generate a random variate $y = \beta(U; u, n-u+1)$, where $U \sim U(0,1)$, and return the quantile $x = F^{-1}(y)$ as the uth order statistic with parent distribution F. Recall that F^{-1} is the inverse distribution function (or quantile function) and is defined by

$$F^{-1}(y) = \inf\{x : F(x) \geq y\}.$$

In the case that F is uniform $[0,1]$, $y = F^{-1}(y)$. Note that if $U_1, U_2, \ldots, U_n$ are i.i.d. $U(0,1)$ random variables and $X_1, X_2, \ldots, X_n$ are i.i.d. random variables with common distribution function F such that $X_i = F^{-1}(U_i)$, then $X_{[i]} = F^{-1}(U_{[i]})$.

4.1 Joint and Conditional Distributions of Order Statistics

The joint density function of $(X_{[u]}, X_{[v]})$, $1 \leq u < v \leq n$ is denoted by $g_{u,v:n}(x, y)$. It can be shown that

$$g_{u,v:n}(x,y) = n! \frac{[F(x)]^{u-1}}{(u-1)!} \frac{[1-F(y)]^{n-v}}{(n-v)!} \frac{[F(y)-F(x)]^{v-u-1}}{(v-u-1)!} f(x)f(y).$$

In particular, the minimum and the maximum, $(X_{[1]}, X_{[n]})$, have the joint density

$$g_{1,n:n}(x,y) = n(n-1)[F(y)-F(x)]^{n-2} f(x)f(y).$$

From the joint distribution of two order statistics we can find the distribution of various other statistics, e.g., the sample range $R = X_{[n]} - X_{[1]}$. The pdf and cdf of the sample range R, respectively, are

$$g_R(R) = n(n-1) \int_{-\infty}^{\infty} [F(x+R) - F(x)]^{n-2} f(x) f(x+R) dx,$$

and

$$G_R(R) = n \int_{-\infty}^{\infty} [F(x+R) - F(x)]^{n-1} f(x) dx.$$

Because $F(x+R)$ may not be expressed as a function of $F(x)$, it often requires numerical integration to solve the integration. In the case that $F = \Phi$ is the standard normal distribution, [Tippett (1925)] showed that the mean value of R is given by

$$E(R) = \int_{-\infty}^{\infty} 1 - [\Phi(x)]^n - [1-\Phi(x)]^n dx.$$

Furthermore, the joint pdf of any number of the order statistics can be constructed, in particular, the joint pdf of all of the order statistics is

$$g_{1,2,\ldots,n:n}(X_{[1]}, X_{[2]}, \ldots, X_{[n]}) = n! f(X_{[1]}) f(X_{[2]}) \ldots f(X_{[n]}).$$

In the case that f is the uniform [0,1] distribution,

$$g_{u,v:n}(x,y) = n! \frac{x^{u-1}}{(u-1)!} \frac{[1-y]^{n-v}}{(n-v)!} \frac{[y-x]^{v-u-1}}{(v-u-1)!},$$

and

$$\text{Cov}[X_{[u]}, X_{[v]}] = \frac{u(n-v+1)}{(n+1)^2(n+2)}.$$

As expected, $\text{Cov}[X_{[u]}, X_{[v]}]$ decreases as the difference between $v - u$ increases, given u or v. Moreover, $\text{Cov}[X_{[1]}, X_{[n]}] \to 0$, as $n \to \infty$. That is, in large samples the minimum and maximum are in general approximately independent. Moreover, the pdf and cdf of the sample range, respectively, are

$$g_R(R) = n(n-1)R^{n-2} \int_0^{1-R} dx = \beta(R; n-1, 2),$$

and

$$G_R(R) = n(1-R)R^{n-1} + R^n.$$

From this we can deduce that

$$E[R] = \frac{n-1}{n+1},$$

and

$$\text{Var}[R] = \frac{2(n-1)}{(n+1)^2(n+2)}.$$

That is, the sample range R approaches the true range and variance of the sample range R becomes smaller as the sample size n becomes larger.

For $1 \leq u < v \leq n$, the conditional distribution of $X_{[u]}$ given $X_{[v]} = x_b$ is the same as the unconditional distribution of the uth order statistic in a sample of size $v-1$ from a new distribution, namely the original F truncated at the right at x_b. In notation,

$$g_{X_{[u]}|X_{[v]}=x_b}(x) =$$

$$\frac{(v-1)!}{(u-1)!(v-1-u)!}(\frac{F(x)}{F(x_a)})^{u-1}(1-\frac{F(x)}{F(x_a)})^{v-1-u}\frac{f(x)}{F(x_a)},$$

$$x < x_b.$$

Similarly, for $1 \leq u < v \leq n$, the conditional distribution of $X_{[v]}$ given $X_{[u]} = x_a$ is the same as the unconditional distribution of the $(v-u)^{th}$ order statistic in a sample of size $n-u$ from a new distribution, namely the original F truncated at the left at x_a. In notation,

$$g_{X_{[v]}|X_{[u]}=x_a}(x) =$$

$$\frac{(n-u)!}{(v-u-1)!(n-v)!}(\frac{F(x)-F(x_a)}{1-F(x_a)})^{v-u-1}(\frac{1-F(x)}{1-F(x_a)})^{n-v}\frac{f(x)}{1-F(x_a)},$$

$$x > x_a.$$

For more detail, see [DasGupta (2011)].

4.2 Using Range Statistics to Perform Equivalence Tests

In this section, we preview the indifference-zone approach and show how range statistics are applied in simulation to perform equivalence tests.

4.2.1 *Indifference-Zone Selection*

Let μ_{i_l} be the lth smallest of the μ_i's, so that $\mu_{i_1} \leq \mu_{i_2} \leq \ldots \leq \mu_{i_k}$. In selection, the goal is to select the best system with the smallest (or largest) expected responses, i.e., system i_1 (or i_k). Let CS denote the event of "correct selection." In a stochastic simulation, a CS can never be guaranteed with certainty. The probability of CS, denoted by P(CS), is a random variable depending on sample sizes and other uncontrollable factors. Moreover, in practice, if the difference between μ_{i_1} and μ_{i_2} is very small, we might not care if we mistakenly choose system i_2, whose expected response is μ_{i_2}. The "practically significant" difference d^* (a positive real number) between a desired and a satisfactory system is called the indifference zone in statistical literature, and it represents the smallest difference which we care about. Therefore, we want a procedure that avoids making a large number of replications or batches to resolve differences less than d^*. That means we want P(CS) $\geq P^*$ provided that $\mu_{i_2} - \mu_{i_1} \geq d^*$, where the minimal CS probability P^* and the "indifference" amount d^* are both specified by the users.

The indifference-zone selection procedure of [Dudewicz and Dalal (1975)] to select the smallest performance measure of k system proceeds as follows.

(1) Simulate the initial n_0 samples for all systems. Compute the first-stage sample means with the equation

$$\bar{X}_i^{(1)}(n_0) = \frac{1}{n_0}\sum_{j=1}^{n_0} X_{ij},$$

and the sample variances with the equation

$$S_i^2(n_0) = \frac{\sum_{j=1}^{n_0}(X_{ij} - \bar{X}_i^{(1)}(n_0))^2}{n_0 - 1},$$

for $i = 1, 2, \ldots, k$.

(2) Compute the required sample sizes with the equation

$$N_i = \max(n_0 + 1, \lceil (h_1 S_i(n_0)/d^*)^2 \rceil), \text{ for } i = 1, 2, \ldots, k, \qquad (4.2)$$

where $\lceil z \rceil$ is the smallest integer that is greater than or equal to the real number z, and h_1 is a critical constant will be described later.

(3) Simulate additional $N_i - n_0$ samples, for $i = 1, 2, \ldots, k$.

(4) Compute the second-stage sample means with the equation

$$\bar{X}_i^{(2)}(N_i - n_0) = \frac{1}{N_i - n_0} \sum_{j=n_0+1}^{N_i} X_{ij}, \text{for } i = 1, 2, \ldots, k.$$

(5) Compute the weighted sample means with the equation

$$\tilde{X}_i(N_i) = W_{i1}\bar{X}_i^{(1)}(n_0) + W_{i2}\bar{X}_i^{(2)}(N_i - n_0), \text{for } i = 1, 2, \ldots, k$$

and select the system with the smallest $\tilde{X}_i(N_i)$.

Note that h_1 (which depends on k, P^*, and n_0) is a constant that can be found from the tables in [Law (2014)] and can be estimated by the procedure of [Chen and Li (2010)]. Furthermore, the weights can be derived from the equations

$$W_{i1} = \frac{n_0}{N_i}\left[1 + \sqrt{1 - \frac{N_i}{n_0}\left(1 - \frac{(N_i - n_0)(d^*)^2}{h^2 S_i^2(n_0)}\right)}\right] \tag{4.3}$$

and $W_{i2} = 1 - W_{i1}$, for $i = 1, 2, \ldots, k$. The expression for W_{i1} was chosen to guarantee $(\tilde{X}_i(N_i) - \mu_i)/(d^*/h_1)$ would have a t distribution with $n_0 - 1$ d.f.; see [Dudewicz and Dalal (1975)].

4.2.2 *Variance of Weighted Sample Means*

Let $\tilde{X}_i$ (i.e., $\tilde{X}_i(N_i)$) be the weighted sample means as defined. The procedure of [Dudewicz and Dalal (1975)] is derived based on the fact that

$$T_i = \frac{\tilde{X}_i - \mu_i}{d^*/h_1},$$

for $i = 1, 2, \cdots, k$ have a t distribution with $n_0 - 1$ d.f. They point out that

$$\frac{\tilde{X}_i - \mu_i}{S_i(n_0)/\sqrt{a_i}} \sim N(0, \sigma_i^2/S_i^2(n_0)),$$

where $a_i = (h_1 S_i(n_0)/d^*)^2$. That means

$$\frac{\tilde{X}_i - \mu_i}{\sigma_i/\sqrt{a_i}} \sim N(0, 1)$$

and $\tilde{X}_i \sim N(\mu_i, \sigma_i^2/a_i)$. Recall that $\bar{X}_i \sim N(\mu_i, \sigma_i^2/N_i)$ and $\sigma_i^2/N_i \leq \sigma_i^2/a_i$ because $a_i \leq N_i = \mathrm{max(n_0 + 1, \lceil (h_1 S_i(n_0)/d^*)^2 \rceil)}$. The difference between

σ_i^2/N_i and σ_i^2/a_i is more apparent when $a_i < n_0$ and the difference increases as $a_i(< n_0)$ becomes smaller. For example, if $a_i = 1$, then $N_i = n_0 + 1$ and the variances of $\bar{X}_i$ and $\tilde{X}_i$ are $\sigma_i^2/(n_0+1)$ and σ^2 respectively. Note that

$$T_i = \frac{\bar{X}_i - \mu_i}{S/\sqrt{N_i}},$$

for $i = 1, 2, \cdots, k$ are t distributed with $N_i - 1$ d.f., hence, they likely have different distributions, i.e., different d.f.

Consequently, if $\mu_i < \mu_j$, then $\Pr[\tilde{X}_i < \tilde{X}_j] \le \Pr[\bar{X}_i < \bar{X}_j]$. Even though $\Pr[\tilde{X}_{i_1} < \tilde{X}_{i_l}$ for $l = 2, 3, \ldots, k] \le \Pr[\bar{X}_{i_1} < \bar{X}_{i_l}$ for $l = 2, 3, \ldots, k]$ has not been proven to be true, we believe it is fairly safe to use $\bar{X}_i$ instead of $\tilde{X}_i$ to perform selection. For a discussion of the performance of using the weighted sample means and the overall sample means, please see [Chen (2011)].

4.2.3 *Effects of the Indifference Amount and Sample Size*

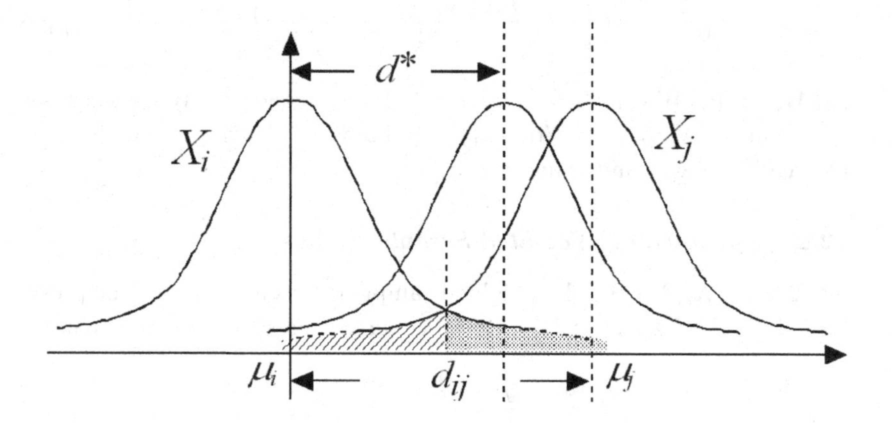

Fig. 4.1 The probability density of performance measures of systems i and j

Traditionally, optimization procedures will resolve to a single selection from all alternatives and hope the selected system is the true best one. With indifference-zone and ordinal optimization, we are selecting a set of good enough alternatives with high probability. One of the tenets of ordinal optimization indicates that it is much easier to determine whether or not alternative i is better than j than to determine the distance $d_{ij} = \mu_j - \mu_i$. Let $\bar{X}_i(N_i)$ and $\bar{X}_j(N_j)$ be the average of N_i and N_j, respectively, normally distributed performance measures of two simulation experiments.

The probability of mis-aligning systems i and j is roughly proportional to the area (of mis-alignment) under the overlapping tail of the two density functions center at μ_i and $\mu_i + d_{ij}$ as illustrated in Figure 4.1, see [Ho (1996)]. Note that $d^* < d_{ij}$ in this case. Moreover, the size of the area is dependent on the variances of $\bar{X}_i(N_i)$ and $\bar{X}_i(N_i)$ as well as the distance d_{ij}. The smaller the variances and the larger the distance, the smaller the area of mis-alignment. For example, the area under the overlapping tail of the two density functions centered at μ_i and $\mu_i + d_{ij}$ is smaller than the center at μ_i and $\mu_i + d^*$.

When we increase the size of the indifference amount d^*, we increase the alignment probability because the area of mis-alignment will be smaller and the number of designs i such that $\mu_i < \mu_{i_1} + d^*$ may increase, i.e., the number of members in the subset G may increase. Consequently, the required sample sizes to achieve the specified P(CS) will be smaller. Hence, increasing the indifference amount can ease the computational burden.

In testing the null hypothesis $\mathrm{H}_0 : \mu_i \leq \mu_{i_1}$, for us to reject the null hypothesis and conclude with confidence level $1 - \alpha$ that $\mu_i > \mu_{i_1}$ is the same as the lower endpoint of the one-tailed $1 - \alpha$ c.i. is positive, i.e., $\bar{X}_i - \bar{X}_{i_1} - w_{ii_1} > 0$, where w_{ii_1} denotes the half-width of the one-tailed $1 - \alpha$ c.i. of $\mu_i - \mu_{i_1}$. The half-width w_{ii_1} depends on the sample sizes and becomes smaller as the sample sizes become large. This implies the sample sizes (N_i and N_{i_1}) should be large enough so that $w_{ii_1} < \bar{X}_i - \bar{X}_{i_1}$. By symmetry of the normal distribution $\Pr[\bar{X}_i - \bar{X}_{i_1} \geq (\mu_i - \mu_{i_1}) - w_{ii_1}] \geq 1 - \alpha$. To obtain $\Pr[\bar{X}_i - \bar{X}_{i_1} > 0] \geq 1 - \alpha$, the sample size should be large enough so that $w_{ii_1} < \mu_i - \mu_{i_1}$.

Let $\hat{d}_{i_1 i_l} = \bar{X}_{i_l} - \bar{X}_{i_1}$. Procedures developed based on the LFC achieve $w_{i_l i_1} < d^*$ and consequently the one-tailed $1 - \alpha$ c.i. of $d_{i_1 i_l}$ $CI1 = (\hat{d}_{i_1 i_l} - d^*, \infty]$. Whereas procedures that take into account sample means attempt to achieve $w_{i_l i_1} < d_{i_1 i_l} = \mu_{i_l} - \mu_{i_1}$ and $CI2 = (\hat{d}_{i_1 i_l} - d_{i_1 i_l}, \infty] \approx (0, \infty]$. Hence, the allocated sample sizes are just large enough for us to conclude $\mu_{i_1} < \mu_i$ (provided $\mu_{i_1} + d^* \leq \mu_i$) with a desired confidence but no more than necessary.

4.2.4 *Equivalence Tests*

This section investigates the hypothesis testing of equivalence of mean of multiple systems using range statistics. The most useful comparison of the system means is done by comparing each system with the unknown best system or best systems within the group. Consider sample means of certain observations drawn at random from k systems. The difference, $R = \bar{X}_{[k]} - \bar{X}_{[1]}$, between the largest and the smallest sample means is called the sample range. The ratio, $q = R/S$, of the range to an independent root-mean-square estimate, S, of the population standard deviation, σ, is called a studentized range. The studentized range is the test statistic of Tukeys range test to find which means are significantly different from one another based on the distribution of q. Note that Tukeys range test assumes equal variances and uses equal sample sizes for each system.

In these procedures, we consider the equivalence tests with critical region $C_R = \{\bar{X}_{[k]} - \bar{X}_{[1]} \leq \delta\}$, where $\delta > 0$ is a user-specified upper bound of the sample range for the hypothesis test. Note that in these cases variances may be unequal among systems and the studentized range can no longer be computed in a straightforward manner.

Let

$$T_i = \frac{\tilde{X}_i - \mu_i}{\delta/h_3} \text{ for } i = 1, 2, \ldots, k.$$

Note that the weight needs to be computed with d^* replaced by δ and the required sample sizes n_i need to be computed by Eq. (4.4), which and the value of the critical constant h_3 will be discussed later. Moreover, T_i's are independent t-distributed random variables with $n_0 - 1$ d.f. Let F denote the cdf of the t distribution with $n_0 - 1$ d.f. Let P(CD) denote the probability of correct decision and let $\tilde{X}_{c_i}$ denote the ith smallest $\tilde{X}_i$. Under the null hypothesis that $\mu_{i_1} = \mu_{i_2} = \cdots = \mu_{i_k}$, we can write

$$\begin{aligned}
\mathrm{P(CD)} &= \Pr[\tilde{X}_{c_k} - \tilde{X}_{c_1} \leq \delta] \\
&= \Pr\left[T_{c_k} - T_{c_1} \leq h_3\right] \\
&= G_R(h_3) \\
&= k \int_{-\infty}^{\infty} [F(t + h_3) - F(t)]^{k-1} f(t) dt.
\end{aligned}$$

The last equality follows because of the cdf of the sample range. We equate the right-hand side to P^* and solve for h_3. Numerical approximation has been used to solve the integration. Nevertheless, other approach can be used to solve h_3, which will be discussed later. Furthermore,

$$\begin{aligned}
\mathrm{P(CD)} &= \Pr[\tilde{X}_{c_k} - \tilde{X}_{c_1} \leq \delta] \\
&\geq \int_{-\infty}^{\infty} [G_{k:k}(t + h_3)]^k g_{1:k}(t) dt \\
&= \int_{-\infty}^{\infty} [F(t + h_3)]^k g_{1:k}(t) dt.
\end{aligned}$$

The inequality follows because T_{c_k} and T_{c_1} are positively correlated.

[Chen (2014a)] proposed using the following procedure to test the null hypothesis that all means are equal.

(1) Simulate the initial n_0 samples for all systems. Compute the first-stage sample means $\bar{X}_i^{(1)}(n_0)$ and the sample variances $S_i^2(n_0)$ for $i = 1, 2, \ldots, k$.
(2) Compute the required sample sizes with the formula

$$N_i = \max(n_0 + 1, \lceil (h_3 S_i(n_0)/\delta)^2 \rceil) \text{ for } i = 1, 2, \ldots, k. \tag{4.4}$$

(3) Simulate additional $N_i - n_0$ samples.
(4) Compute the second-stage sample means with the formula

$$\bar{X}_i^{(2)} = \frac{1}{N_i - n_0} \sum_{j=n_0+1}^{N_i} X_{ij} \text{ for } i = 1, 2, \ldots, k.$$

(5) Compute the weighted sample means $\tilde{X}_i(N_i)$. Reject the null hypothesis that all means are equal when $\tilde{X}_{c_k} - \tilde{X}_{c_1} > \delta$; otherwise do not reject the null hypothesis.

4.2.5 *Confidence Interval Half Width of Interest*

Let $\vartheta = T_{c_k} - T_{c_1}$. Then the cdf of ϑ

$$G_R(h_3) = \Pr[\vartheta \le h_3].$$

Note that the cdf $G_R(h_3)$ is determined only by the d.f. of the t-distribution given k. By definition h_3 is the P^* quantile of the distribution of ϑ when $G_R(h_3) = P^*$.

Furthermore,

$$\begin{aligned}\vartheta &= T_{c_k} - T_{c_1} \\ &= \frac{(\tilde{X}_{c_k} - \tilde{X}_{c_1}) - (\mu_{c_k} - \mu_{c_1})}{\delta/h_3} \\ &= \frac{(\tilde{X}_{c_k} - \tilde{X}_{c_1}) - (\mu_{c_k} - \mu_{c_1})}{\sqrt{S_{c_k}^2(n_0)/(2N_{c_k}) + S_{c_1}^2(n_0)/(2N_{c_1})}}.\end{aligned}$$

Without loss of generality, we temporarily assume that N_{c_k} and N_{c_1} are real numbers. The last equality holds because

$$\frac{S_{c_k}^2(n_0)}{2N_{c_k}} = \frac{S_{c_1}^2(n_0)}{2N_{c_1}} = \frac{\delta^2}{2h_3^2} \text{ and } \sqrt{\frac{S_{c_k}^2(n_0)}{2N_{c_k}} + \frac{S_{c_1}^2(n_0)}{2N_{c_1}}} = \frac{\delta}{h_3}.$$

Consequently,

$$G_R(h_3) = \Pr\left[\frac{(\tilde{X}_{c_k} - \tilde{X}_{c_1}) - (\mu_{c_k} - \mu_{c_1})}{\sqrt{S^2_{c_k}(n_0)/(2N_{c_k}) + S^2_{c_1}(n_0)/(2N_{c_1})}} \le h_3\right] = P^*.$$

Let

$$w_{c_k c_1} = \frac{h_3}{\sqrt{2}}\sqrt{\frac{S^2_{c_k}(n_0)}{N_{c_k}} + \frac{S^2_{c_1}(n_0)}{N_{c_1}}}.$$

Then

$$\Pr[\tilde{X}_{c_k} - \tilde{X}_{c_1} - w_{c_k c_1} \le \mu_{c_k} - \mu_{c_1}] = P^*.$$

By definition, $w_{c_k c_1}$ is the one-tailed P^* confidence interval half width.Under the null hypothesis that all means are equal, $\mu_{c_k} - \mu_{c_1} = 0$. Consequently,

$$\Pr[\tilde{X}_{c_k} - \tilde{X}_{c_1} \le w_{c_k c_1} \le \delta] = P^*.$$

To obtain $w_{c_k c_1} \le \delta$, sample sizes $N_i \ge (h_3 S_i(n_0)/\delta)^2$, provided $(h_3 S_i(n_0)/\delta)^2 > n_0$.

The expected value of $\tilde{X}_{c_j} - \tilde{X}_{c_i}$ (the difference of the weighted sample means of systems c_j and c_i) decreases as $j - i > 0$ decreases. Even though the variance of $\tilde{X}_{c_j} - \tilde{X}_{c_i}$ increases as $j - i > 0$ decreases, the P^* quantile of the random variate $\tilde{X}_{c_j} - \tilde{X}_{c_i}$ is smaller than h_3 for $i, j = 1, 2, \ldots, k-1$, $i \ne j$. Consequently, let

$$w_{ij} = \frac{h_3}{\sqrt{2}}\sqrt{\frac{S_i^2(n_0)}{N_i} + \frac{S_j^2(n_0)}{N_j}},$$

then

$$\Pr[\tilde{X}_i - \tilde{X}_j - w_{ij} \le \mu_i - \mu_j] \ge P^* \text{ for } i, j = 1, 2, \cdots, k, i \ne j$$

and

$$\Pr[\tilde{X}_i - \tilde{X}_j - w_{ij} \le \mu_i - \mu_j \le \tilde{X}_i - \tilde{X}_j + w_{ij}] \ge 2P^* - 1$$

$$\text{for } i, j = 1, 2, \cdots, k, i \ne j.$$

4.3 Statistical Analysis of the Range

In this section, we investigate how to simulate sample range and estimate the values of the cdf of the range.

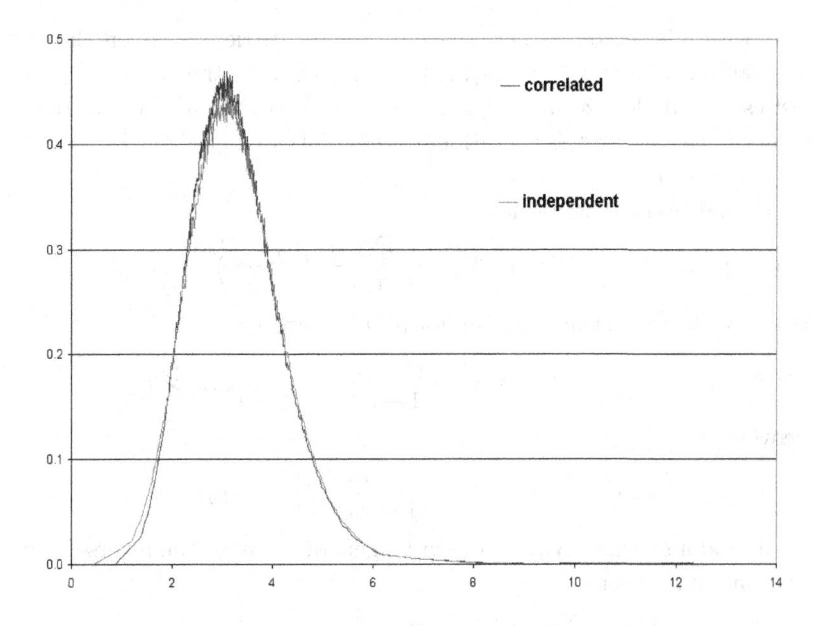

Fig. 4.2 Empirical probability density of a range

4.3.1 *Simulating the Sample Range*

Recall that a straightforward way of simulating sample range is to generate n pseudorandom samples from the distribution $F(X)$ and record the sample minimum and maximum. The sample range can then be calculated with the sample minimum and maximum. This general method requires generating n samples. On the other hand, the sample minimum and maximum can also be generated directly with the properties of order statistics. Let $U_1 = \beta(u_1, 1, n)$ and $U_2 = \beta(u_2, n, n)$, where $u_1, u_2 \sim U(0,1)$ and are independent. Then $F^{-1}(U_1)$ and $F^{-1}(U_1)$ are, respectively, the sample minimum and maximum. However, the distribution of the random variate $F^{-1}(U_2) - F^{-1}(U_1)$ will have greater variance than the true distribution of the sample range because the minimum and maximum of samples are not independent.

Figure 4.2 lists the empirical distributions of the range of 10 t distributed (with 19 d.f.) random variables. The correlated and independent distributions are, respectively, the distribution of $t_{[10]} - t_{[1]}$ when $t_{[1]}$ and $t_{[10]}$ are generated correlated and independently. The correlated one has a smaller variance, the value of the mode is smaller and the pdf value of the mode is greater, hence, the higher quaniles (say $p > 0.5$) will be smaller.

From the discussion earlier, the nth order statistic $x_{[n]}$ given the first order statistic $x_{[1]} = x_a = \beta(u_1, 1, n)$ is the same as the $(n-1)^{th}$ order statistics of sample size $n-1$ with the same distribution F truncated at x_a. Hence, we can use the conditional probability of $x_{[n]}$ given $x_{[1]} = x_a$ to estimate the range.

The conditional distribution

$$g_{X_{[n]}|X_{[1]}=x_a}(x) = (n-1)(\frac{F(x) - F(x_a)}{1 - F(x_a)})^{n-2}\frac{f(x)}{1 - F(x_a)}, x > x_a.$$

In the case that F is the cdf of uniform [0,1] distribution

$$g_{X_{[n]}|X_{[1]}=x_a}(x) = (n-1)(\frac{x - x_a}{1 - x_a})^{n-2}\frac{1}{1 - x_a}, x > x_a.$$

Moreover,

$$G_{X_{[n]}|X_{[1]}=x_a}(x) = (\frac{x - x_a}{1 - x_a})^{n-1}, x > x_a.$$

The random variate $X_{[n]}$ can then be generated using the inverse transformation, i.e., we set

$$(\frac{X_{[n]} - x_a}{1 - x_a})^{n-1} = U \sim U(0,1),$$

to obtain

$$X_{[n]} = x_a + (1 - x_a) * U^{1/(n-1)}.$$

The sample range $R = X_{[n]} - X_{[1]}$. In cases that F is not uniform [0,1], the sample range $R = F^{-1}(X_{[n]}) - F^{-1}(X_{[1]})$. Recall that if $X_i = F^{-1}(U_i)$, then $X_{[i]} = F^{-1}(U_{[i]})$.

4.3.2 *Estimating Quantiles of the Range*

The equivalence tests are derived from the equation

$$\mathrm{P(CD)} = \Pr[T_{c_k} - T_{c_1} \leq h_3].$$

We equal the right-hand side to P^* to obtain h_3. Note that by definition h_3 is the P^* quantile of the variate $T_{c_k} - T_{c_1}$. In this section, we present an approach to estimate the quantiles of a range.

[Chen and Kelton (2008)] control the precision of quantile estimates by ensuring that the p quantile estimator $\hat{x}_p$ satisfies the following:

$$\Pr[x_p \in \hat{x}_{p\pm\epsilon}] \geq 1-\alpha_1, \text{ or equivalently } \Pr[|F(\hat{x}_p)-p| \leq \epsilon] \geq 1-\alpha_1. \quad (4.5)$$

Figure 4.3 demonstrates this requirement with an exponential distribution, see [Eickoff et al. (2006)] Using this precision requirement (i.e., Eq. (4.5)),

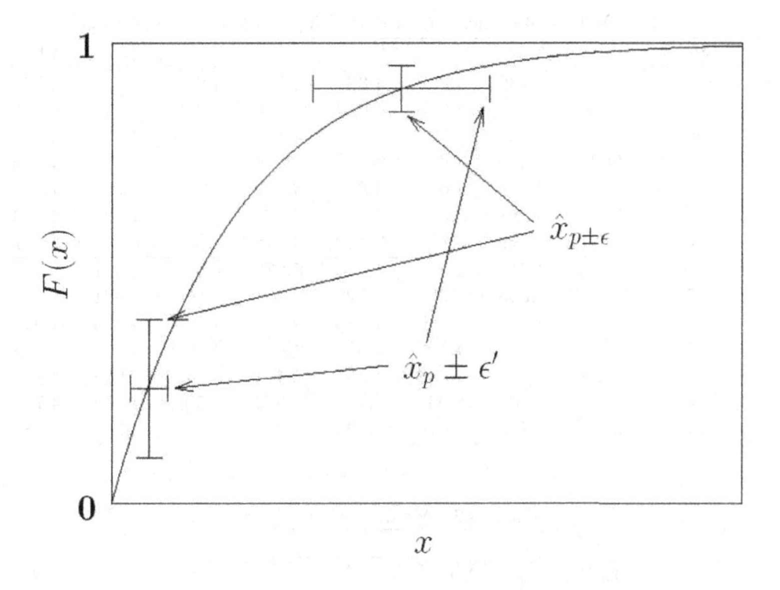

Fig. 4.3 Confidence intervals for quantiles

the required sample size n_p for a fixed-sample-size procedure of estimating the p quantile of an i.i.d. sequence is the minimum n_p that satisfies

$$n_p \geq \frac{z^2_{1-\alpha_1/2}p(1-p)}{\epsilon^2}, \tag{4.6}$$

where $z_{1-\alpha_1/2}$ is the $1-\alpha_1/2$ quantile of the standard normal distribution, ϵ is the maximum proportional half-width of the confidence interval, and $1-\alpha_1$ is the confidence level. For example, if the data are independent and one wants to have 95% confidence that the coverage of the 0.9 quantile estimator has no more than $\epsilon = 0.0001$ (coverage) deviation from the true but unknown quantile, the required sample size is $n_p \geq 34574400$ (i.e., $1.960^2 0.9(1-0.9)/0.0001^2$). Consequently, we will have 97.5% confidence that the quantile estimate will cover at least $p - 0.0001$ (for $p \geq 0.9$), with sample size 34574400.

To avoid storing and sorting the entire sequence, a modified version of the histogram-approximation procedure of [Chen and Kelton (2008)] was used to estimate quantiles. See [Chen and Li (2010)] for more details. Table 4.1 lists the critical constant h_3.

Table 4.1 Values of critical constant h_3 (Quantile of sample range)

P^*	n_0/k	3	4	5	6	7	8	9	10
0.90	10	3.310	3.739	4.053	4.301	4.507	4.683	4.837	4.974
	15	3.150	3.542	3.825	4.046	4.228	4.382	4.516	4.634
	20	3.081	3.457	3.727	3.937	4.108	4.253	4.379	4.489
	25	3.042	3.409	3.672	3.876	4.041	4.182	4.303	4.409
	30	3.017	3.379	3.637	3.837	3.999	4.136	4.254	4.358
	35	2.999	3.358	3.613	3.810	3.970	4.105	4.221	4.323
	40	2.987	3.342	3.595	3.790	3.948	4.082	4.196	4.297
0.95	10	3.874	4.304	4.618	4.869	5.076	4.693	5.411	5.549
	15	3.650	4.033	4.311	4.528	4.707	4.859	4.991	5.108
	20	3.554	3.919	4.180	4.385	4.551	4.744	4.816	4.924
	25	3.501	3.855	4.108	4.305	4.466	4.601	4.719	4.822
	30	3.467	3.815	4.062	4.255	4.411	4.544	4.658	4.758
	35	3.444	3.787	4.031	4.221	4.374	4.504	4.616	4.714
	40	3.427	3.766	4.008	4.195	4.347	4.475	4.585	4.682

Table 4.2 Mean configuration

Model	μ_i
EMC	$\mu_i = 0$, for $i = 1, \ldots, k$
LFC	$\mu_1 = 0$ and $\mu_i = d^*$, for $i = 2, 3, \ldots, k$

4.4 Empirical Experiments

In this section, we present some empirical results. Instead of using stochastic system simulation examples, which offer less control over the factors that affect the performance of a procedure, we use various normally distributed random variables to represent the systems. We chose the first-stage sample size to be $n_0 = 20$. The number of systems under consideration is $k = 10$. The critical range δ is set to $0.5/\sqrt{n_0}$ and $d^* = 2\delta$. The required minimal P(CS), P^*, is set to 0.90.

The equal means configuration (EMC) and the least favorable configuration (LFC) are used. Under the LFC, $\mu_1 + d^* = \mu_i$, for $i \neq 1$. Tables 4.2 and 4.3 list the configurations of mean and variance, respectively. With the EMC, a correct decision means we do not reject the null hypothesis $H_0 : \mu_1 = \mu_2 = \cdots = \mu_k$. With the LFC, a correct decision means we reject the null hypothesis.

We performed 10,000 independent runs to estimate the observed P(CD), which is computed by dividing the number of times the procedure made a CS by 10,000. We list the results (the observed P(CD), the average sample size of each simulation run $\overline{T}$, i.e., $\overline{T} = \sum_{r=1}^{10000} \sum_{i=1}^{k} N_{r,i}/10000$, $N_{r,i}$ is the total number of replications or batches for system i in the rth independent

Table 4.3 Variance configuration

Model	μ_i
Equal	1
Increasing	$1 + (i-1)d^*$
Decreasing	$\frac{1}{1+(i-1)d^*}$

Table 4.4 The observed P(CD) with $P^* = 0.90$, $n_0 = 20$ and equal variances among systems

Setting/k	3	4	5	6	7	8	9	10
EMC	0.9270	0.9329	0.9333	0.9427	0.9453	0.9466	0.9523	0.9545
LFC	1.0	1.0	1.0	1.0	1.0	1.0	1.0	1.0
$\overline{T}$	2274	3820	5550	7435	9442	11570	13799	16113

Table 4.5 The observed P(CD) with $P^* = 0.90$, $n_0 = 20$ and increasing variances among systems

Setting/k	3	4	5	6	7	8	9	10
EMC	0.9237	0.9305	0.9342	0.9429	0.9427	0.9480	0.9506	0.9516
LFC	1.0	1.0	1.0	1.0	1.0	1.0	1.0	1.0
$\overline{T}$	2784	5103	8037	11596	15785	20632	26150	32337

Table 4.6 The observed P(CD) with $P^* = 0.90$, $n_0 = 20$ and decreasing variances among systems

Setting/k	3	4	5	6	7	8	9	10
EMC	0.9273	0.9321	0.9353	0.9410	0.9414	0.9458	0.9493	0.9475
LFC	1.0	1.0	1.0	1.0	1.0	1.0	1.0	1.0
$\overline{T}$	1901	2965	4032	5084	6112	7116	8093	9039

run) in Tables 4.4 through 4.6. All the observed P(CD)s are greater than the specified nominal level of 0.90. We believe it is because the critical constant h_3 is a round up of the true value. Hence, as the number of systems k increases the P(CD)s becomes larger. The procedure correctly increases (decreases) the sample sizes as the variances increase (decrease). Consequently, the observed P(CD)s do not degrade as the variances increase (decrease).

4.5 Summary

Order statistics have proved useful for both theoretical and computational purposes. Important special cases of order statistics are the minimum and maximum value of samples, and the sample median and other sample quantiles. We have investigated several properties of order statistics and discussed how order statistics are applied in simulation, e.g., generate the distributions of order statistics without sorting and presented simulation procedures that are derived based on the inference of order statistics, e.g., hypothesis tests of equivalence of means. Properties of order statistics allow us to derive many simulation algorithms in a concise manner. These procedures are easy to understand and simple to implement.

Chapter 5

Order Statistics from Correlated Normal Random Variables

In this chapter, we investigate an important special case of order statistics, namely, the underlying variates are correlated normal random variables and show how it is related to ranking and selection. [Gupta et al. (1973)] point out that the study of these order statistics is important because that the joint distributions of standardized correlated variables often tend asymptotically to the multivariate normal distribution. [Arellano and Genton (2007, 2008)] have derived the density of correlated normal random variables from a multivariate normal distribution. [Liu et al. (2012)] develop a method to construct exact simultaneous confidence interval for a finite set of contrasts of three, four or five generally correlated normal means.

We are interested in the case that $X_1, X_2, \cdots, X_n$ are n standardized normal random variables with correlation matrix $\{\rho_{ij}\}$. In the special case that $\rho_{ij} = \rho \geq 0$ $(i \neq j)$, it can be shown that

$$F_n(h; \rho) = \Pr[X_{[n]} \leq h] = \int_{-\infty}^{\infty} \Phi^n\{(x\rho^{1/2} + h)/(1-\rho)^{1/2}\}\phi(x)dx, \quad (5.1)$$

where $\Phi(x)$ and $\phi(x)$ are, respectively, the cumulative distribution function and the probability density function of a standardized normal random variable, see [Dunnett and Sobel (1955)]. By the symmetry of the normal distribution $F_n(h; \rho)$ is also equal to $\Pr[X_{[1]} \geq -h]$.

[Gupta et al. (1973)] point out that in several cases of applications, the percentage points (quantile) of the distribution of $X_{[n]}$ are needed and provide tables of percentage points for several selected values of n and positive ρ.

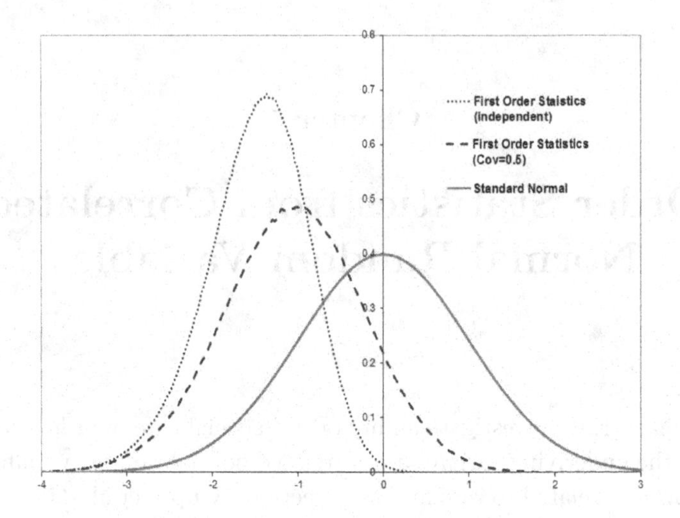

Fig. 5.1 Empirical density of the first-order statistic of 9 $N(0, 1)$ random variables

5.1 Order Statistics of Correlated Random Variables

[Chen (2014b)] investigated the distribution of the order statistic $X_{[u]}$ when X are correlated. Figure 5.1 shows the empirical density functions of three first-order statistic of 9 standard normal random variables: 1) samples are independent; 2) samples with covariance of 0.5; 3) samples with covariance of 1. See [Arellano and Genton (2007, 2008)] for more graphical illustrations of distributions of order statistics of correlated normal random variables. The standard normal can be reviewed as the first-order statistic of 9 perfectly correlated standard normal random variables. [Gupta et al. (1973)] show that the cdf of $X_{[u]}$ when X are correlated normal random variables with $\rho_{ij} = \rho\ (i \neq j)$ is:

$$F_u(h_a;\rho) = \Pr[X_{[u]} \leq h_a] = \int_{-\infty}^{\infty} I_{q(z)}(u, n-u+1)d\phi(z),$$

where $q(z) = \Phi\{(z\rho^{1/2} + h_a)/(1-\rho)^{1/2}\}$, and $I_q(a,b)(a,b > 0)$ is the usual incomplete beta function. When the correlation matrix has the structure $\rho_{ij} = c_i c_j (i \neq j)$, where $-1 \leq c_i \leq 1$ for $i = 1, 2, \ldots, n$, then $F_u(h_a;\rho)$ with $u = n$ is reduced to Eq. (5.1). Furthermore, X_i can be represented as

$$X_i = c_i Z_0 + (1 - c_i^2)^{1/2} Z_i, \text{ for } i = 1, 2, \ldots, n, \tag{5.2}$$

where $Z_0, Z_1, \ldots, Z_n$ are independent standard normal random variables. If $\rho_{ij} = \rho \geq 0$, then the structural assumption that $\rho_{ij} = c_i c_j$ is satisfied

Table 5.1 Values of critical constant h_a with $\rho = 0.5$ (known equal variances)

n/α	0.010	0.025	0.050	0.100	0.250
2	2.5581	2.2124	1.9165	1.5770	1.0139
3	2.6852	2.3492	2.0622	1.7336	1.1895
4	2.7718	2.4420	2.1605	1.8385	1.3056
5	2.8372	2.5117	2.2341	1.9164	1.3912
6	2.8894	2.5671	2.2924	1.9781	1.4586
7	2.9327	2.6132	2.3407	2.0289	1.5139
8	2.9701	2.6523	2.3816	2.0721	1.5605
9	3.0022	2.6865	2.4173	2.1095	1.6008
10	3.0307	2.7165	2.4486	2.1424	1.6362
12	3.0795	2.7677	2.5019	2.1981	1.6959
14	3.1198	2.8101	2.5461	2.2441	1.7452
16	3.1545	2.8464	2.5837	2.2833	1.7868
18	3.1846	2.8778	2.6163	2.3172	1.8230
20	3.2115	2.9058	2.6452	2.3473	1.8547
22	3.2352	2.9306	2.6709	2.3740	1.8830
24	3.2571	2.9532	2.6943	2.3981	1.9085
26	3.2767	2.9737	2.7155	2.4202	1.9318
28	3.2949	2.9926	2.7349	2.4403	1.9530
30	3.3117	3.0100	2.7529	2.4589	1.9726
32	3.3270	3.0262	2.7697	2.4763	1.9908
34	3.3417	3.0414	2.7853	2.4923	2.0077
36	3.3554	3.0555	2.7998	2.5074	2.0235
38	3.3684	3.0690	2.8137	2.5216	2.0384
40	3.3803	3.0815	2.8266	2.5350	2.0524
42	3.3921	3.0935	2.8389	2.5476	2.0657
44	3.4029	3.1046	2.8504	2.5595	2.0782
46	3.4133	3.1155	2.8615	2.5710	2.0901
48	3.4232	3.1258	2.8721	2.5818	2.1015
50	3.4328	3.1357	2.8823	2.5923	2.1123

by taking $c_i = \sqrt{\rho}$ for all i. An observation of $X_{[n]}$ from correlated normal random variables with correlation ρ can be simulated as follows. Let $y = \beta(U; n, 1)$ and set $X_{[n]} = \sqrt{\rho}Z_0 + \sqrt{1-\rho}F^{-1}(y)$.

5.1.1 *Method of Evaluation of the Percentage Points*

Let α be the probability of committing the Type I error, i.e., do not accept the null hypothesis when it is correct. The value of h_a is the $1-\alpha$ quantile of the correlated normal random variable $X_{[n]}$. We used the method described in Chapter 6 to estimate the percentage points (i.e., quantiles). This approach is flexible and can be used to estimate the critical constant for the problem at hand, even when the correlations are unknown or unequal.

Table 5.1 lists the critical constant h_a with $\rho = 0.5$ for selected α and

Table 5.2 Values of critical constant h_b with $\rho = 0.5$ (unknown equal variances with the number of observations $n_0 = 20$)

n/α	0.010	0.025	0.050	0.100	0.250
2	2.8416	2.4018	2.0458	1.6562	1.0436
3	3.0143	2.5760	2.2222	1.8364	1.2326
4	3.1353	2.6973	2.3444	1.9600	1.3601
5	3.2290	2.7903	2.4373	2.0534	1.4553
6	3.3052	2.8657	2.5125	2.1284	1.5312
7	3.3691	2.9289	2.5753	2.1910	1.5940
8	3.4239	2.9832	2.6292	2.2447	1.6476
9	3.4727	3.0312	2.6766	2.2916	1.6941
10	3.5161	3.0737	2.7186	2.3332	1.7355

sample sizes n. The values are within ± 0.0004 of those listed in [Gupta et al. (1973)], where a wide range of critical constant is listed. When the variance is unknown, it needs to be estimated by sample variance. Each sample mean will be the average of n_0 observations, the standard error of these n_0 observations are used as an estimator of standard deviation. Table 5.2 in Section 5.2.3 lists the critical constant h_b with $\rho = 0.5$ and the number of observations $n_0 = 20$ for selected sample sizes n.

5.2 Applications of Correlated Order Statistics

[Gupta et al. (1973)] discuss several applications illustrating the use of the tables. These applications relate to multiple decision procedures, multiple comparison problems and some tests of hypotheses. In this section, we review multiple comparison and multiple decision procedures. Multiple comparisons provide simultaneous confidence intervals on selected differences among the systems. Multiple-decision procedure is used to select the system with the smallest or the largest mean.

5.2.1 *Multiple Comparisons with a Control*

In multiple comparisons with a control, we are interested in the lower confidence limits for $\mu_i - \mu_0$ $(i = 1, 2, \ldots, k)$ with joint confidence P^* in the case of $k+1$ normal populations π_i $(i = 0, 1, \ldots, k)$ with mean μ_i $(i = 0, 1, \ldots, k)$ and common known variance σ^2. Let X_{ij} denote the jth observation of the ith system. Let $\bar{X}_0 = \sum_{j=1}^{m_0} X_{0j}/m_0$ and $\bar{X}_i = \sum_{j=1}^{n_0} X_{ij}/n_0$ $(i = 1, 2, \ldots, k)$ be the sample means based on a sample size m_0 from the control π_0 and samples of size n_0 from each of the k populations π_i

$(i = 1, 2, \ldots, k)$. Then $\bar{X}_i - \bar{X}_0 \sim N(\mu_i - \mu_0, \sigma^2(1/m_0 + 1/n_0))$, where $N(\mu, \sigma^2)$ denotes the normal distribution with mean μ and variance σ^2. Let h denote the P confidence critical constant. By the definition of the confidence limit

$$\Pr[\mu_i - \mu_0 \geq \bar{X}_i - \bar{X}_0 - h\sigma\sqrt{1/m_0 + 1/n_0}] = P \text{ for } i = 1, 2, \ldots, k.$$

We are interested in constant h_a such that

$$\Pr[\mu_i - \mu_0 \geq \bar{X}_i - \bar{X}_0 - h_a\sigma\sqrt{1/m_0 + 1/n_0} \text{ for } i = 1, 2, \ldots, k] = P^*.$$

Let

$$Z_i = \frac{(\bar{X}_i - \bar{X}_0) - (\mu_i - \mu_0)}{\sigma\sqrt{1/m_0 + 1/n_0}} \text{ for } i = 1, 2, \ldots, k.$$

Then

$$\Pr[Z_i \leq h_a \text{ for } i = 1, 2, \ldots, k] = P^*.$$

Furthermore, let $Z_{[k]} = \max_{i=1}^{k} Z_i$ for $i = 1, 2, \ldots, k$. Then

$$\Pr[Z_{[k]} \leq h_a] = P^*.$$

Note that Z_i are standardized normal variables with equal correlation $\rho = n_0/(m_0+n_0)$ and $Z_{[k]}$ is the kth order statistic of Z_i. Consequently, the constant $h_a = H_a(P^*, k, n_0/(m_0+n_0))$, where $H_a(1-\alpha, k, \rho)$ is the percentage point with Type I error probability α, sample size k, and correlation ρ.

Note that $\mathbf{Z} = (Z_1, \ldots, Z_k)^T$ has an exchangeable multivariate normal distribution (i.e., its covariance matrix is equicorrelated) with a common mean $\mu = 0$, a common variance σ^2 and a common correlation coefficient ρ. That is, $\mathbf{Z}$ is multivariate normal $N_k((\mu - \mu_0)\mathbf{1}_k, \sigma^2\{(1-\rho)I_k + \rho\mathbf{1}_k\mathbf{1}_k^T\})$ with $\rho \in [0, 1)$, $\mathbf{1}_k \in R^k$ a vector of ones, and $I_k \in R^{k\times k}$ the identity matrix. It can be shown that the probability density function $f_{z_{[k]}}$ of $z_{[k]}$ is

$$f_{z_{[k]}}(z) = k\phi_1(z; \mu, \sigma^2)\Phi_{k-1}(\sqrt{1-\rho}(z-\mu)/\sigma\mathbf{1}_{k-1}; \mathbf{0}, I_{k-1} + \rho\mathbf{1}_{k-1}\mathbf{1}_{k-1}^T),$$

where $z \in R$, $\phi_1(z; \mu, \sigma^2)$ is the marginal probability density function of Z_k and $\Phi_{k-1}(\mathbf{z}; \mu, \Sigma)$ is the $k-1$ multivariate cumulative distribution function. See [Arellano and Genton (2008)].

5.2.2 *Multiple Decision (Ranking and Selection)*

In multiple decision, we are interested in selecting the system having the largest mean, with known equal variance σ^2 under the indifference zone setting. Without loss of generality, assume that $\mu_i = \mu_k - d^*$ for $i = 1, 2, \cdots, k-1$. Here d^* is an indifference amount specified by the users. The probability of correct selection, P(CS), is then

$$\begin{aligned} \mathrm{P(CS)} &= \Pr[\bar{X}_i < \bar{X}_k \text{ for } i = 1, 2, \cdots, k-1] \\ &= \Pr[\frac{\bar{X}_i - \mu_i}{\sigma/\sqrt{n_0}} < \frac{\bar{X}_k - \mu_k}{\sigma/\sqrt{n_0}} + \frac{d^*}{\sigma/\sqrt{n_0}} \text{ for } i = 1, 2, \cdots, k-1] \\ &= \Pr[Z_{[k-1]} < Z_k + \frac{d^*}{\sigma/\sqrt{n_0}}] \\ &= \int_{-\infty}^{\infty} \Phi^{k-1}(z + \frac{d^*}{\sigma/\sqrt{n_0}}) d\Phi(z). \end{aligned}$$

We set the right-hand side to P^* to solve for n_0. Note that

$$(Z_1 - Z_k, \ldots, Z_{k-1} - Z_k) \sim N_{k-1}(-d^* \mathbf{1}_{k-1}, \sigma^2\{0.5 I_{k-1} + 0.5 \mathbf{1}_{k-1}\mathbf{1}_{k-1}^T\}).$$

Because $\Pr[(Z_{[k-1]} - Z_k)/\sqrt{2} < h_a] \geq P^*$, (here $h_a = H_a(P^*, k-1, 0.5)$), the sample size n_0 should large enough such that $\sqrt{2} h_a \leq d^* \sqrt{n_0}/\sigma$, i.e., $n_0 \geq 2(h_a \sigma / d^*)^2$. See [Bechhofer (1954)] for details.

5.2.3 *Multiple Comparisons with a Control: Unknown Equal Variances*

Let μ_0 denote the performance measure of the control (i.e., a benchmark system). We are interested in the lower confidence limits for $\mu_i - \mu_0$ ($i = 1, 2, \ldots, k$) with joint confidence P^* in the case of $k+1$ normal populations π_i ($i = 0, 1, \ldots, k$) with mean μ_i ($i = 0, 1, \ldots, k$) and common unknown variance σ^2. Let $\bar{X}_i$ ($i = 0, 1, \ldots, k$) be the sample means based on samples of size n_0 from each of the $k+1$ populations π_i ($i = 0, 1, \ldots, k$). Then $\bar{X}_i - \bar{X}_0 \sim N(\mu_i - \mu_0, \sigma^2 2/n_0)$. Let h_b denote the critical constant and S^2 be the unbiased variance estimate of the unknown variance σ^2. We are interested in constant h_b such that

$$\Pr[\mu_i - \mu_0 \geq \bar{X}_i - \bar{X}_0 - h_b S\sqrt{2/n_0} \text{ for } i = 1, 2, \ldots, k] = P^*.$$

Let

$$T_i = \frac{(\bar{X}_i - \bar{X}_0) - (\mu_i - \mu_0)}{S\sqrt{2/n_0}} \text{ for } i = 1, 2, \ldots, k.$$

Then

$$\Pr[T_i \le h_b \text{ for } i = 1, 2, \ldots, k] = P^*.$$

Furthermore, let $T_{[k]} = \max_{i=1}^k T_i$. Then

$$\Pr[T_{[k]} \le h_b] = P^*.$$

Note that T_i for $i = 1, 2, \ldots, k$ are t-distributed (with $n_0 - 1$ d.f.) with equal correlation $\rho = 1/2$ and $T_{[k]}$ is the kth order statistic of T_i. That is, $\mathbf{T} = (T_1, \ldots, T_k)^T$ is an exchangeable random vector with a multivariate Student t distribution with $v = n_0 - 1$ d.f. and

$$\mathbf{T} \sim \text{Student}_k(\mu \mathbf{1}_k, \sigma^2\{(1-\rho)I_k + \rho \mathbf{1}_k \mathbf{1}_k^T\}, v).$$

It can be shown that the pdf $f_{T_{[k]}}$ of $T_{[k]}$ is

$$f_{T_{[k]}}(t) = k t_1(t; \mu, \sigma^2, v) \times$$

$$T_{k-1}(\sqrt{1-\rho}(t-\mu)/\sigma \mathbf{1}_{k-1}; \mathbf{0}, \frac{v+z^2}{v+1}\{I_{k-1} + \rho \mathbf{1}_{k-1}\mathbf{1}_{k-1}^T\}, v+1), t \in R,$$

where $t_1(t; \mu, \sigma^2)$ is the marginal probability density function of T_k, $z^2 = (t-\mu)^2/\sigma^2$ and $T_{k-1}(\mathbf{z}; \mu, \Sigma, v+1)$ is the $k-1$ multivariate t cumulative distribution function. See [Arellano and Genton (2008)]. Consequently, the constant $h_b = H_b(P^*, k, 0.5, n_0)$, where $H_b(1-\alpha, k, \rho, n_0)$ is the percentage point with Type I error probability α, sample size k, correlation ρ, and the number of observations n_0.

We compute the sample variances with the equation

$$S_i'^2 = \frac{\sum_{j=1}^n ((X_{ij} - X_{0j}) - (\bar{X}_i - \bar{X}_0))^2}{n_0 - 1},$$

for $i = 1, 2, \ldots, k$. The quantity $S_i'/\sqrt{n_0}$ will be used in the place of $S\sqrt{2/n_0}$ for each T_i for $i = 1, 2, \ldots, k$. Note that S_i' already incorporates the standard error of X_0, hence, the multiplier $\sqrt{2}$ is no longer needed.

5.2.4 *Multiple Comparisons with a Control: Unknown Unequal Variances*

Let $\tilde{X}_i$ be the weighted sample means (with the first-stage sample size n_0) as defined in [Dudewicz and Dalal (1975)]. Then

$$\vartheta_i = T_i - T_0 = \frac{(\tilde{X}_i - \tilde{X}_0) - (\mu_i - \mu_0)}{d^*/h_c} = \frac{(\tilde{X}_i - \tilde{X}_0) - (\mu_i - \mu_0)}{\sqrt{S_i^2(n_0)/(2N_i) + S_0^2(n_0)/(2N_0)}}.$$

Here $S_i^2(n_0)$ is the unbiased estimator of the variance σ_i^2 with n_0 samples. The correlations of $\tilde{X}_i - \tilde{X}_0$ and $\tilde{X}_j - \tilde{X}_0$ for $i \neq j$ are unknown, but the correlations of ϑ_i (i.e., $T_i - T_0$) and ϑ_j (i.e., $T_j - T_0$) for $i \neq j$ are $\rho_{ij} = 1/2$. Without loss of generality, we temporarily assume that N_i and N_0 are real numbers. The last equality holds because

$$\frac{S_i^2(n_0)}{2N_i} = \frac{S_0^2(n_0)}{2N_0} = \frac{(d^*)^2}{2h_c^2} \text{ and } \sqrt{\frac{S_i^2(n_0)}{2N_i} + \frac{S_0^2(n_0)}{2N_0}} = \frac{d^*}{h_c}.$$

The distribution of ϑ_i is symmetric and its percentile (or quantile) can be evaluated numerically. Let

$$w_{i0} = \frac{h_c}{\sqrt{2}}\sqrt{\frac{S_i^2(n_0)}{N_i} + \frac{S_0^2(n_0)}{N_0}}.$$

Then $w_{i0} = d^*$. In practice, N_i for $i = 0, 1, \ldots, k$ are integers and $w_{i0} \leq d^*$. For some discussion of the properties of $\bar{X}$ and $\tilde{X}$, see [Chen (2011)].

We are interested in the value h_c such that $\Pr[\tilde{X}_i - \tilde{X}_0 - w_{i0} \leq \mu_i - \mu_0$ for $i = 1, 2, \ldots, k] = P^*$. That is, $\Pr[\vartheta_i \leq h_c$ for $i = 1, 2, \ldots, k] = P^*$. Similarly, $\Pr[T_{[k]} - T_0 \leq h_c] = \int_{-\infty}^{\infty} G_{k:k}(t + h_c) f(t) dt = P^*$. Note that T_i for $i = 0, 1, \ldots, k$ are independent and the parent distribution of $G_{k:k}$ is the t-distribution with $n_0 - 1$ d.f. With the indifference-zone approach, the allocated sample sizes obtain $w_{i0} \leq d^*$. To increase the efficiency, the strategy of [Chen (2004)] is to allocate sample sizes such that $w_{i0} \leq \mu_i - \mu_0$. Because μ_i for $i = 0, 1, 2, \ldots, k$ are unknown, in practice, the procedure obtains $w_{i0} \leq \bar{X}_i - \bar{X}_0$.

5.3 Empirical Experiments

In this section, we present some empirical results of comparison with a control. The number of systems under consideration is $k = 5$.

5.3.1 *Experiment 1: Known Equal Variances*

In this experiment, the unknown true means for all systems are $\mu_i = 0$ and the known variances of all systems are $\sigma_i^2 = 1.0$ for $i = 0, 1, 2, \ldots, k$. The sample size for each system is $n_0 = 20$. Consequently, the correlation of $\bar{X}_i - \bar{X}_0$ for $i = 1, 2, \ldots, k$ is $\rho = 0.5$. The nominal confidence level is $P^* = 0.90, 0.95$, and 0.99, hence, the critical constant $h_a(P^*, 5, 0.5) = 1.9164, 2.2341$, and 2.8372. We list $\hat{P}(C)$, the percentage

Table 5.3 Percentage of the coverage with $k = 5$ and known variances

P^*	0.90	0.95	0.99
$\hat{P}(C)$	0.8982	0.9505	0.9889

Table 5.4 Percentage of the coverage with $k = 5$ and unknown variances

P^*	0.90	0.95	0.99
$\hat{P}(C)$	0.8999	0.9508	0.9891

that the simultaneous lower confidence limits contain the true differences $\mu_i - \mu_0$ (i.e., $0 \geq \bar{X}_i - \bar{X}_0 - h_a\sqrt{2}/\sqrt{20}$) for $i = 1, 2, \ldots, k$. Table 5.3 lists the results of experiment 1. The observed coverages are close to the nominal values.

5.3.2 *Experiment 2: Unknown Equal Variances*

In this experiment, the unknown means for all systems are $\mu_i = 0$ and the unknown variances of all systems are $\sigma_i^2 = 1.0$ for $i = 0, 1, 2, \ldots, k$. Note that the variances of $\bar{X}_i - \bar{X}_0$ need to be estimated. The sample size for each system is $n_0 = 20$. The nominal confidence level is $P^* = 0.90, 0.95$, and 0.99, hence, the critical constant $h_b(P^*, 5, 0.5, 20) = 2.0534, 2.4373$, and 3.2290. We list $\hat{P}(C)$, the percentage that the simultaneous lower confidence limits contain the true differences $\mu_i - \mu_0$ (i.e., $0 \geq \bar{X}_i - \bar{X}_0 - h_b S_i'/\sqrt{20}$) for $i = 1, 2, \ldots, k$. Table 5.4 lists the results of experiment 2. The observed coverages are close to the nominal values. The coverages in experiments 1 and 2 are basically the same. However, the half width of the c.i's in experiment 2 are wider than those in experiment 1 because the variance needs to be estimated using sample variance.

5.3.3 *Experiment 3: Unknown Unequal Variances*

In this experiment, the unknown means for all systems are $\mu_i = 0$ and the unknown variances of system i σ_i^2 for $i = 0, 1, 2, \ldots, k$ may be different. Table 5.5 lists the configuration of variances. The required parameter d^* is set to 0.5 or 1.0, which implies that the targeted one-tailed confidence interval half-width is 0.5 or 1.0, respectively. The sample size for each system is computed according to

$$N_i = \max(n_0 + 1, \lceil (h_c S_i(n_0)/d^*)^2 \rceil), \text{ for } i = 0, 1, \ldots, k. \quad (5.3)$$

Table 5.5 Variance configuration

Model	$\sigma_0^2, \sigma_1^2, \sigma_2^2, \sigma_3^2, \sigma_4^2, \sigma_5^2$
Equal	1, 1, 1, 1, 1, 1
Increasing	$1, 1+d^*, 1+2d^*, 1+3d^*, 1+4d^*, 1+5d^*$
Decreasing	$1, \frac{1}{1+d^*}, \frac{1}{1+2d^*}, \frac{1}{1+3d^*}, \frac{1}{1+4d^*}, \frac{1}{1+5d^*}$

Table 5.6 Percentage of the coverage with equal means and $d^* = 0.5$

Configuration	P^*	half width	0.90	0.95	0.99
	Weighted $\hat{P}(C)$	d^*	0.8936	0.9474	0.9905
	Overall $\hat{P}(C)$	d^*	0.9011	0.9499	0.9904
Equal	Weighted $\hat{P}(C)$	w_{i0}	0.8842	0.9442	0.9903
	Overall $\hat{P}(C)$	w_{i0}	0.8932	0.9470	0.9903
	$\bar{T}$		200	276	462
	Weighted $\hat{P}(C)$	d^*	0.9027	0.9475	0.9902
	Overall $\hat{P}(C)$	d^*	0.9072	0.9497	0.9904
Increasing	Weighted $\hat{P}(C)$	w_{i0}	0.8985	0.9457	0.9899
	Overall $\hat{P}(C)$	w_{i0}	0.9031	0.9481	0.9903
	$\bar{T}$		446	618	1041
	Weighted $\hat{P}(C)$	d^*	0.8952	0.9504	0.9900
	Overall $\hat{P}(C)$	d^*	0.9471	0.9690	0.9917
Decreasing	Weighted $\hat{P}(C)$	w_{i0}	0.8422	0.9247	0.9865
	Overall $\hat{P}(C)$	w_{i0}	0.9151	0.9558	0.9903
	$\bar{T}$		142	167	250

The nominal confidence level is $P^* = 0.90, 0.95$, and 0.99, hence, the critical constant $h_c(P^*, 5, 20) = 2.8703, 3.3767$, and 4.3872. Note that $h_c(P^*, k, n_0) = h_1(P^*, k+1, n_0)$. We list the percentage that the simultaneous lower confidence limits (constructed four different ways) contain the true differences $\mu_i - \mu_0$ (i.e., $0 \geq \tilde{X}_i - \tilde{X}_0 - d^*$, $0 \geq \bar{X}_i - \bar{X}_0 - d^*$, $0 \geq \tilde{X}_i - \tilde{X}_0 - w_{i0}$, $0 \geq \bar{X}_i - \bar{X}_0 - w_{i0}$) for $i = 1, 2, \ldots, k$.

Tables 5.6 and 5.7 list the results of experiment 3 with $d^* = 0.5$ and 1.0, respectively. In addition to the observed coverages, we also list the average sample size of each simulation run $\overline{T}$, i.e., $\overline{T} = \sum_{r=1}^{10000} \sum_{i=0}^{k} N_{r,i}/10000$, $N_{r,i}$ is the total number of replications or batches for system i in the rth independent run. The procedure correctly increases and decreases the allocated sample sizes as the variance increases and decreases, respectively. The observed coverages of the c.i. constructed by $0 \geq \tilde{X}_i - \tilde{X}_0 - d^*$ and $0 \geq \bar{X}_i - \bar{X}_0 - w_{i0}$ are close to the nominal values. Because the overall sample mean $\bar{X}_i$ has smaller variance than the weighted sample mean $\tilde{X}_i$, the coverages of the confidence intervals built by $\bar{X}_i$ is slightly greater than those built by $\tilde{X}_i$, with the same half width.

Table 5.7 Percentage of the coverage with equal means and $d^* = 1.0$

Configuration	P^*	half width	0.90	0.95	0.99
	Weighted $\hat{P}(C)$	d^*	0.9026	0.9478	0.9905
	Overall $\hat{P}(C)$	d^*	0.9973	0.9973	0.9983
Equal	Weighted $\hat{P}(C)$	w_{i0}	0.6780	0.8195	0.9704
	Overall $\hat{P}(C)$	w_{i0}	0.9145	0.9546	0.9927
	$\bar{T}$		126	126	132
	Weighted $\hat{P}(C)$	d^*	0.9018	0.9455	0.9907
	Overall $\hat{P}(C)$	d^*	0.9618	0.9778	0.9948
Increasing	Weighted $\hat{P}(C)$	w_{i0}	0.8008	0.8962	0.9835
	Overall $\hat{P}(C)$	w_{i0}	0.8876	0.9445	0.9915
	$\bar{T}$		193	253	408
	Weighted $\hat{P}(C)$	d^*	0.8997	0.9484	0.9896
	Overall $\hat{P}(C)$	d^*	0.9997	1.0000	0.9999
Decreasing	Weighted $\hat{P}(C)$	w_{i0}	0.5797	0.7257	0.9215
	Overall $\hat{P}(C)$	w_{i0}	0.9310	0.9672	0.9922
	$\bar{T}$		126	126	127

In the setting that $d^* = 1.0$, the initial sample size (i.e., 126= (20+1)6) is large enough to achieve greater precision than the specified precision in many cases. The weighted sample means purposely lose information so that the observed coverages of the c.i. constructed by $\tilde{X}_i - \tilde{X}_0 - d^*$ are close to the nominal values. On the other hand, in those cases the observed coverages of the c.i. constructed by $\tilde{X}_i - \tilde{X}_0 - w_{i0}$ are less than the nominal values because $N_i = n_0 + 1$ for $i = 0, 1, \ldots, k$ are greater than the required sample sizes (say a_i for $i = 0, 1, \ldots, k$). Hence,

$$w_{i0} = \frac{h_c}{\sqrt{2}}\sqrt{\frac{S_i^2(n_0)}{N_i} + \frac{S_0^2(n_0)}{N_0}} \leq \frac{h_c}{\sqrt{2}}\sqrt{\frac{S_i^2(n_0)}{a_i} + \frac{S_0^2(n_0)}{a_0}} \approx d^*.$$

That is, in the cases that $a_i \leq n_0$, the variance of $\tilde{X}_i$ is greater than $\bar{X}_i$ while the c.i. half width w_{i0} is computed based on the variance of $\bar{X}_i$ and results in low coverages. On the other hand, the coverages of c.i. constructed by $\bar{X}_i - \bar{X}_0 - d^*$ are greater than the nominal values. i.e., the half width is wider than necessary.

5.4 Summary

We have investigated properties of order statistics of correlated normal random variables and discussed how order statistics are applied in simulation, e.g., to perform comparison with a control and multiple decision. Furthermore, we proposed a new approach to estimate the values of the cdf of correlated normal random variables. The new approach is flexible and can be used to estimate the critical constants for the problem at hand, even when the correlations are unknown or unequal.

Chapter 6

Histogram and Quasi-Independent Procedure

Simulation is often used to investigate system characteristics, i.e., the distributional properties of an output statistic calculated from the simulation's results, such as the mean or variance of system performance measures. In many applications we want to estimate the probability that a future observation will exceed a given level during some specified epoch. For example, a machine may break down if a certain temperature is reached. Among many other aspects, reliability analysis (risk analysis) studies the expected life and the failure rate of a component or a system of components linked together in some structure. It is often of interest to estimate the reliability of the component/system from the observed lifetime data. With probability modeling, normal and lognormal densities are commonly used to model certain lifetimes in reliability and survival analysis as well as risk management.

Risk management is the identification, assessment, and prioritization of risks to minimize the probability of unfortunate events, e.g., machine break down, loss of financial assets, cyberattacks. Inadequate risk management can result in severe consequences for individuals as well as the entire society. For instance, the U.S. subprime mortgage crisis and the associated recession were largely caused by the loose credit risk management of financial firms. Risk management often starts with a probabilistic risk assessment, a systematic and comprehensive methodology to identify and evaluate risks. Based on the assessment, the planner then comes up a strategy and/or plan to minimize the loss (cost), or to maximize the return (opportunity). The simplest models (of analyzing risks) often consist of a probability multiplied by an impact (severity of the possible adverse consequences). Understanding risks may be difficult as multiple factors (impacts) can contribute to the total probability of risk. In financial mathematics and financial risk management, value at risk (a quantile-based risk measurement) is a widely used risk measure of the risk of loss on specific portfolio of financial assets.

We propose a method to construct an empirical distribution of the output statistic of interest. The associated distributional characteristics are then estimated based on the empirical distribution. For $0 < p < 1$, the p *quantile* (or $100p$ percentile) of a distribution is the value at or below which $100p$ percent of the distribution lies. Related to quantiles, a *histogram* is a graphical estimate of the underlying probability density (or mass) function. The range of the output data is divided into intervals, or bins, and the number or proportion of the observations falling into each bin is tabulated or plotted. A histogram can be constructed with a properly selected set of quantiles.

We extend the *quasi-independent* (QI) algorithm of [Chen and Kelton (2003)] to determine the simulation run length and use grid points to construct a histogram (with a selected set of quantiles) and use the histogram as an empirical distribution for estimating quantiles. [Iglehart (1976); Seila (1982); Hurley (1995)] have developed quantile-estimation algorithms based on grid points. However, their procedures require that users enter the values of the grid points. Moreover, the procedures of [Iglehart (1976)] and [Seila (1982)] require that the underlying processes have the regenerative property, which can be difficult to know in large, complex simulation models. The procedures of [Heidelberger (1984)] and [Raatikainen (1990)] requires that the output sequences satisfy ϕ*-mixing* conditions. Similarly, the procedure described in this chapter requires that as well. There are many other quantile-estimation procedures, see, e.g., [Alexopoulos et al. (2014)] and references therein.

The asymptotic validity of the QI procedure occurs as the QI sequence appears to be independent, as determined by the *runs* test. The main advantage of the approach is that by using grids to approximate the underlying distribution, it avoids the burden of storing and sorting all the observations, which becomes prohibitive in very long runs that might be needed in steady-state simulation. However, the savings come at a cost. Using interpolation to obtain quantile estimates introduces bias. Fortunately, the bias can be reduced by specifying finer grid points, which would then require longer execution time, so there is a natural trade-off here.

6.1 Introduction and Definitions

Sample quantiles based on the order statistics are the natural estimator of quantiles and have strong theoretical basis. Furthermore, order-statistics quantile estimates are non-parametric and are valid regardless the shape of the underlying distribution. On the contrary, indirect approaches or parametric quantile estimation procedures, which assume that the data are drawn from a known parametric family of distributions, need to be used with caution.

For example, if one *approximates* the steady-state waiting time in system distribution of the M/M/1 queuing process with the service rate $\nu = 1$ and the arrival rate $\lambda = 0.8$ with a normal distribution and uses the sample mean $\hat{\mu}$ plus $z_{0.9}S_x$ to estimate the 0.9 quantile will get good results by coincidence. Here $z_{1-\alpha}$ is the $1-\alpha$ quantile of the standard normal distribution and S_x is the standard deviation of the samples. The high accuracy of estimating the 0.9 quantiles does not hold when estimating some other quantiles.

Let W_i denote the waiting time in system of the ith customer and let $\rho = \lambda/\nu$ be the traffic intensity. Then, if $\rho < 1$, the theoretical steady-state waiting time in system distribution of this M/M/1 queuing process is

$$F(x) = \Pr[W_i \leq x] \rightarrow 1 - e^{-(\nu-\lambda)x} \text{ as } i \rightarrow \infty$$

for all $x \geq 0$. Hence, the true steady-state distribution and the true 0.9 quantile of this M/M/1 queuing process are, respectively, $F(x) \rightarrow 1 - e^{-(1-0.8)x}$ and approximately 11.512925. Furthermore, both the mean and standard deviation of this distribution are 5 (i.e., $1/(\nu-\lambda) = 1/(1-0.8)$). Consequently, the naive quantile estimator is approximately 11.41 (i.e., $5 + 1.282 \times 5$), which is close to the true 0.9 quantile. However, this naive quantile estimator may be away from the true quantile when $p \neq 0.9$.

Let MM1(0.8) denote the steady-state waiting time in system distribution of the M/M/1 queuing process with traffic intensity 0.8 and let $N(\mu, \sigma^2)$ denote the normal distribution with mean μ and variance σ^2. Figure 6.1 compares the true distributions of $N(5, 5^2)$ and MM1(0.8) for $x \geq 0$. From the diagram, the naive estimator will be close to the true quantile around the 0.23 and 0.89 quantiles, where these two distributions intersect. The naive estimator is likely to be biased high for $0.23 < p < 0.89$. For instance, the true 0.5 quantile of $N(5, 5^2)$ and MM1(0.8) are, respectively, 5 and 3.465736. Again, one may get good results by coincide when the mean or variance estimators are biased low, which are likely when the underlying sequences are correlated.

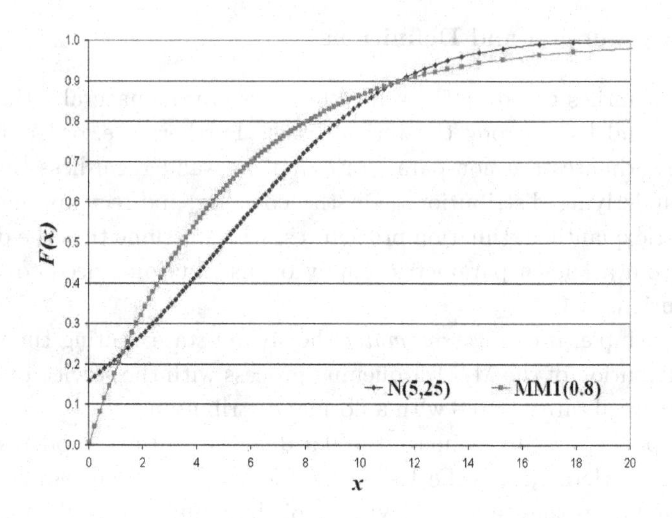

Fig. 6.1 True distributions of $N(5, 5^2)$ and MM1(0.8)

It turns out that the diagram will be similar if the value of ρ is changed, i.e., the two distributions intersect around the 0.23 and 0.89 quantiles regardless the value of ρ. Let p quantile be an intersect of the two distributions. Then $\hat{x}_p = 1/(\nu - \lambda) + z_p/(\nu - \lambda)$ and

$$F(\hat{x}_p) = 1 - e^{-(\nu-\lambda)\hat{x}} = 1 - e^{-(1+z_p)}.$$

Note that $F(\hat{x}_p)$ is independent of ν and λ. Consequently, we will have good results using normal approximation to estimate p quantiles (for p close to 0.23 or 0.89) of the steady-state waiting time distribution of the M/M/1 queuing process regardless the value of ρ; assuming the mean and variance estimates are accurate. Unfortunately, in most cases the distribution from which quantiles are to be estimated is unknown and we don't know whether or where it intersects with the approximated (normal) distribution or how close these two distributions are. Furthermore, it is not known whether the intersects stay at the same quantiles when the parameters of the system change.

Moreover, non-order-statistics (indirect) quantile estimates have not shown to be normally distributed and can not be used directly as input of selection procedures. *Batch means* can be used to *manufacture* approximately i.i.d. normal data so that they can be used as input of selection procedures. Nevertheless, incorporating batch means into the process introduces additional complexities. Note that if samples are i.i.d. normal, the c.i. coverage should stay the same when the sample sizes are fixed regardless whether batch means are used.

On the other hand, order-statistics (direct) quantile estimates are asymptotically normally distributed and can be used as input of selection procedures directly. Consequently, any ranking and selection procedures that are developed to select the design with the best mean can be used to select design with the best quantile. The criticism of the direct approach are that they require large sample sizes, large data storage, and are computationally intensive; especially when estimating multiple quantiles simultaneously. With the advance of modern computers, except in extreme cases most of these issues are no longer major concerns. Moreover, the histogram-approximation-quantile-estimation procedure does not require storing and sorting all the observations and can estimate multiple quantiles simultaneously without specifying the quantiles to be estimated in advance.

6.1.1 *The Natural Estimators*

Let $F(\cdot)$ and $F_N(\cdot)$ respectively denote the true and the (sampled) empirical steady-state cumulative distribution function of the simulation-output process under study, where N is the simulation run length (we assume a discrete-time output process $X_1, X_2, \ldots$). For purpose of analysis, it is convenient to express $F_N(\cdot)$ as

$$F_N(x) = \frac{1}{N}\sum_{i=1}^{N} I_{(-\infty,x]}(X_i), \text{ where } I_{(-\infty,x]}(X_i) = \begin{cases} 1 \text{ if } X_i \le x, \\ 0 \text{ if } X_i > x. \end{cases}$$

In addition to the autocorrelations of the stochastic process output sequence approaching zero as the lag between observations increases, we require two things for the method to work. First, $F_N(\cdot)$ must converge to $F(\cdot)$ as $N \to \infty$. In random (i.e., independent) sampling, $X_1, X_2, \ldots, X_N$ is a random sample of size N, and if $N \to \infty$, then $F_N(\cdot)$ will tend with probability one to $F(\cdot)$. Second, if a statistic T is estimating some characteristic property Ψ of the distribution, then this characteristic must satisfy certain smoothness properties (e.g., continuity and differentiability). Most characteristics such as moments and percentiles are smooth and so their estimators have distributions that can be estimated. The procedure may fail when T is a statistic estimating a function that is not smooth. The property Ψ could be, for example, the mean, variance, or a quantile. The natural point estimator for Ψ, denoted by $\hat{\Psi}$, is typically the sample mean, the sample variance, the sample quantile, or a simple function of the relevant order statistics, chosen to imitate the performance measure Ψ. Furthermore, the natural estimators are appropriate for estimating any Ψ, even in the presence of autocorrelation, which follows since $F_N(\cdot)$ converges to $F(\cdot)$.

6.1.2 *Proportion Estimation*

Here, we are interested in estimating the probability p that the output random variable X_i belongs to a pre-specified set ω: $p = \Pr[X_i \in \omega]$. Let I_i be the indicator functions

$$I_i = \begin{cases} 1 \text{ if } X_i \in \omega, \\ 0 \text{ otherwise.} \end{cases}$$

An estimate of p is based on a transformation of the output sequence $\{X_i\}$ to the sequence $\{I_i\}$, $i = 1, 2, \ldots, N$:

$$\hat{p} = \frac{1}{N} \sum_{i=1}^{N} I_i.$$

For data that are i.i.d., the following properties of I_i are well known [Hogg et al. (2012)]: $E(I_i) = p$ and $\text{Var}(I_i) = p(1-p)$. Moreover, $E(\hat{p}) = p$ and $\text{Var}(\hat{p}) = p(1-p)/N$. Thus, an exact c.i. for the estimated proportion $\hat{p}$ can be obtained using the binomial distribution. That is,

$$p \in \hat{p} \pm t_{n-1,1-\alpha/2} \sqrt{\frac{\hat{p}(1-\hat{p})}{n-1}}.$$

However, we cannot assume that the I_i's are independent. Instead, we assume merely that the sequence $\{I_i\}$ is covariance-stationary. A discrete-time stochastic process $X_1, X_2, \ldots$ is said to be covariance-stationary if

$$\mu_i = \mu \text{ and } \sigma_i^2 = \sigma^2 \text{ for } i = 1, 2, \ldots.$$

Here $-\infty < \mu < \infty$ and $\sigma^2 < \infty$. Furthermore, $\gamma_{i,i+k} = \text{Cov}(X_i, X_{i+k})$ is independent of i for $k = 1, 2, \ldots$. That is, the covariance between X_i and X_{i+k} depends only on k. Hence, the covariance-stationary lag-k covariance of the process $\gamma_k = \gamma_{i,i+k}$ for all i.

The steady-state variance constant of the process is

$$\text{SSVC}(I_i) = \gamma_0 + 2 \lim_{N \to \infty} \sum_{k=1}^{N-1} (1 - k/N)\gamma_k. \tag{6.1}$$

Note that the lag-k correlation of the process

$$C_k = \frac{\gamma_{i,i+k}}{\sqrt{\sigma_i^2 \sigma_{i+k}^2}} = \frac{\gamma_k}{\sigma^2} = \frac{\gamma_k}{\gamma_0} \text{ for } k = 0, 1, 2, \ldots.$$

It follows from the definition of the steady-state variance constant that for large N

$$\text{Var}(\hat{p}) \approx \text{SSVC}(I_j)/N. \tag{6.2}$$

Since $\hat{p}$ is based on the mean of the random variable I_j, we can use any method developed for estimating the variance of the mean to estimate $\text{Var}(\hat{p})$.

6.1.3 *Quantile Estimation*

If $\{X_i : i = 1, 2, \ldots, N\}$ is a sequence of i.i.d. random variables from a continuous distribution $F(x)$ with pdf $f(x)$, let x_p $(0 < p < 1)$ denote the $100p^{th}$ percentile (or the p quantile), which has the property that $F(x_p) = \Pr[X \leq x_p] = p$. Thus, $x_p = \inf\{x : F(x) \geq p\}$. Note that estimating quantiles is the inverse of of the problem of estimating a proportion or probability. If $\{Y_i : i = 1, 2, \ldots, N\}$, are the order statistics corresponding to the X_i's from N independent observations, (i.e. Y_i is the ith smallest of $X_1, X_2, \ldots, X_N$) then a point estimator for x_p based on the order statistics is the sample p quantile $\hat{x}_p$,

$$\hat{x}_p = Y_{\lceil Np \rceil}, \tag{6.3}$$

where $\lceil z \rceil$ denotes the integer ceiling (round-up function) of the real number z.

If the X_i's are i.i.d., we have the following properties [David and Nagaraja (2003)]:

$$E(\hat{x}_p) = x_p - \frac{p(1-p)f'(x_p)}{2(n+2)f^3(x_p)} + O(1/N^2),$$

$$\text{Var}(\hat{x}_p) = \frac{p(1-p)}{(n+2)f^2(x_p)} + O(1/N^2). \tag{6.4}$$

It follows from the central limit theorem for the sample mean that

$$\frac{\hat{x}_p - x_p}{\sqrt{\text{Var}(\hat{x}_p)}} \xrightarrow{D} N(0,1) \text{ as } N \to \infty. \tag{6.5}$$

If the X_i's are correlated, however (as is usually the case if they are a simulation-run output sequence), quantile estimation is much more difficult than in the i.i.d. case. The usual order-statistic point estimate $\hat{x}_p$

is still asymptotically unbiased; however, its variance needs to be adjusted for the autocorrelation of the sequence. [Sen (1972)] shows that for ϕ-mixing sequences, the variance of sample quantiles is inflated by a factor of $\mathrm{SSVC}(I_j)/(p(1-p))$.

6.2 Methodologies

This section presents the methodologies we will use for the quantile and histogram estimation.

6.2.1 *Determining the Simulation Run Length*

The procedure will progressively increase the simulation run length N until a subsequence of n observations (taken from the original output sequence) appears to be independent, as determined by the runs test.

Let

$$[P]_0^1 = \begin{cases} P \text{ if } 0 \leq P \leq 1, \\ 0 \text{ if } P < 0, \\ 1 \text{ if } P > 1, \end{cases}$$

and let ϵ be the desired half-width of the level $1-\alpha_1$ confidence interval for the proportion. Note that parameters ϵ and α_1 are associated with the proportion and are different from ϵ', ϵ'', and α that will be introduced later. The half-width ϵ is dimensionless; it is a proportion value with no measurement unit and must be between 0 and $\max(p, 1-p)$, $0 < p < 1$.

Recall that if the data are i.i.d. and we would like to have 95% confidence that the 0.5 quantile estimator has no more than $\epsilon = 0.005$ (coverage) deviation from the true but unknown quantile, the required simulation run length (sample size) is $N = n_{0.5} \geq \frac{1.96^2 0.5(1-0.5)}{0.005^2} = 38,416$, see Chapter 4. For correlated sequence, the simulation run length will be $N = n_{0.5}l$, where l is the lag that systematic samples pass the test of independence, see Chapter 2. The rational is that the sample size N determined this way is sufficiently large such that quantile estimates are i.i.d. normal.

6.2.2 *Histogram Approximation*

To avoid storing and sorting the whole output sequence, the procedure computes sample quantiles only at certain grid points and use (four-point) Lagrange interpolation [Knuth (1998)] to estimate the general p quantile.

The procedure requires that users enter the values of ϵ and α_1 for the desired precision and confidence. There are two categories of grids: main grids and auxiliary grids. Main grids are constructed based on the initial observations that "anchor" the grid of the simulation-generated histogram, while auxiliary grids are extensions of main grids to ensure the grids cover all future observations. The number of main grid points is $G_m = \lceil 1/\epsilon \rceil$, and the number of auxiliary grid points is $G_a = 2\lceil \delta G_m \rceil + 3$, where $0 < \delta < 1$. Based on previous experiments, we recommend $\delta \geq 0.1$. The total number of grid points is thus $G = G_m + G_a$. Let the beginning and ending indices of the main grid points be respectively $b = \lceil \delta G_m \rceil + 1$ and $e = b + G_m$; denote the main grid points by g_i for $i = b+1, \ldots, e$. The value of the grid points $g_0, g_1, \ldots, g_{G-1}$ will be constructed as follows: g_0 and g_{G-1} are set to $-\infty$ and ∞ (in practice the minimum and maximum values on the host computer) respectively. If the analyst knows what can be the minimum or maximum values of the distribution, those values should be used instead. For example, the waiting time in any queuing system cannot be negative, so the analyst should enter 0 as the minimum g_0.

The selection of grid size is based on initial observations. Grid point g_b is set to the minimum of the initial n, $2n$, or $3n$ observations, depending on the degree of the autocorrelation of the sequence, as determined by the runs test. Grid points g_{b+i}, $i = 1, 2, \ldots, G_m$, are set to the i/G_m quantile of the initial n, $2n$, or $3n$ observations, depending on the degree of autocorrelation of the sequence. We will set grid points g_1 through g_{b-1} and g_{e+1} through g_{G-2} to appropriate values so that g_1 through g_{b+1} will have the same grid size (the length between two adjacent grid points) and g_{e-1} through g_{G-2} will have the same grid size. A corresponding array $\eta_1, \eta_2, \ldots, \eta_{G-1}$ is used to store the number of observations between two consecutive grid points. For example, the number of observations between g_{i-1} and g_i is η_i.

Once the QI algorithm has determined that the simulation run length is long enough for the required precision, we can then compute the quantile estimator by Lagrange interpolation of the quantile at four grid points. The array $\eta_i, i = 1, 2, \ldots, G-1$, stores the number of observations between grid points g_{i-1} and g_i, so the quantile of g_i at grid point i can be computed as $\hat{F}(g_i) = p_i = \sum_{j=1}^{i} n_j/N$, for $i = 1, 2, \ldots, G-1$, where $N = \sum_{j=1}^{G-1} n_j$ is the simulation run length (i.e., number of observations). Thus, for some k such that $p_{k-1} < p \leq p_k$, the p quantile estimator can be computed as follows. Let

$$\varpi_j = \prod_{i=1, i\neq j}^{4} \frac{p - p_{k+i-3}}{p_{k+j-3} - p_{k+i-3}}, \text{ for } j = 1, 2, 3, 4,$$

and

$$\hat{x}_p = \sum_{j=1}^{4} \varpi_j g_{k+j-3}. \tag{6.6}$$

In two extreme cases, $p_0 < p \leq p_1$ or $p_{G-2} < p \leq p_{G-1}$, linear interpolation will be used. Users can also rerun the simulation with finer grid points around the estimated quantile value to get a more accurate estimate.

Because we are estimating quantiles of stochastic systems, inferences based on only one output sequence are unreliable. Therefore, we will run R (we use $R = 3$ in the algorithm) replications to get R quantile estimators. Let $\hat{x}_{p,r}$ denote the estimator of x_p in the rth replication. We use

$$\bar{\hat{x}}_p = \frac{1}{R} \sum_{r=1}^{R} \hat{x}_{p,r} \tag{6.7}$$

as a point estimator of x_p. Assuming that the asymptotic approximation is valid with the simulation run length determined by the procedure, each $\hat{x}_{p,r}$ has a limiting normal distribution. By the central limit theorem, a c.i. for x_p using the i.i.d. $\hat{x}_{p,r}$'s can be approximated using standard statistical procedures. That is, the ratio

$$T = \frac{\bar{\hat{x}}_p - x_p}{S/\sqrt{R}}$$

would have an approximate t distribution with $R - 1$ d.f., where

$$S^2 = \frac{1}{R-1} \sum_{r=1}^{R} (\hat{x}_{p,r} - \bar{\hat{x}}_p)^2$$

is the usual estimator of $\sigma_p^2(n)$, the variance of $\hat{x}_p$. This would then lead to the $100(1 - \alpha)\%$ c.i., for x_p,

$$\bar{\hat{x}}_p \pm t_{R-1,1-\alpha/2} \frac{S}{\sqrt{R}}. \tag{6.8}$$

The quasi-independent quantile estimation algorithm:

(1) The quantity ϵ is the desired half-width of the $100(1-\alpha_1)\%$ c.i. for the proportion specified by the user, B_s is the size of the buffer that is used to store QI samples, and r is the number of iterations. Each iteration r contains two sub-iterations r_A and r_B. Note that the loop of steps 3 through 10 is considered an iteration.
(2) Compute the required QI sequence size n from Eq. (4.6) with the user-specified α_1, and p set to 0.5. If $n < 4000$, set $n = 4000$. If n is odd, then increase n by 1. Set $B_s = 3n$ and $r = 0$. Generate $N = n$ observations as the initial samples.
(3) Carry out runs tests to determine whether the sequence appears to be independent. The runs test uses 4000 samples. If this is the initial iteration, use lag-1 samples in the QI sequence. If this is a r_A iteration, use lag-2 samples in the QI sequence. If this is a r_B iteration, use lag-3 samples in the QI sequence.
(4) If the QI samples appear to be independent or the run length has reached B_s, compute the value of grid points as discussed, i.e., the grids between two consecutive main grid points will contain a proportion approximately equal to ϵ of the distribution.
(5) If the QI samples appear to be independent, go to step 11.
(6) If the current iteration is a r_A iteration, start a r_B iteration. If the current iteration is the initial one or a r_B iteration, set $r = r + 1$ and start a r_A iteration.
(7) If this is the 1_A^{th} or 1_B^{th} iteration, generate n observations, store those values in the buffer after the ones already there, and go to step 3.
(8) If this is a r_A iteration $(r \geq 2)$, then discard every other sample in the buffer, and reindex the rest of the $3n$ samples in the first half of the buffer. Generate $2^{r-2}n$ observations and store those $n/2$ samples that are lag-2^{r-1} observations apart in the later portion of the buffer.
(9) If this is a r_B iteration $(r \geq 2)$, generate $2^{r-1}n$ observations and store those n samples that are lag-2^{r-1} observations apart in the later portion of the buffer.
(10) Go to step 3.
(11) Compute the p quantile estimator according to Eq. (6.6).
(12) Run R replications and compute the confidence interval of the quantile estimator according to Eq. (6.8).
(13) Let ϵ' be the desired absolute half-width (or let $\gamma|\hat{x}_p|$ be the desired relative half-width). If the absolute (or relative) half-width of the c.i.

is less than ϵ' (or $\gamma|\hat{x}_p|$), terminate the algorithm. Note that ϵ' has the same units of measurement as do the observations and does not need to be a fraction.

(14) Run one more replication, and set $R = R + 1$. Go to step 13.

The QI procedure addresses the problem of determining the simulation run length that is required to satisfy the assumptions of independence and normality of the quantile estimator. Theoretically, if these assumptions are satisfied, then the actual coverage of the c.i.'s should be close to the pre-specified level. However, we are not sure whether the asymptotic approximation is valid, so the c.i. constructed by Eq. (6.8) may have coverage less than specified. On the other hand, the quantile estimators should satisfy the precision requirement of Eq. (4.5).

Let the half-width be $w = t_{R-1,1-\alpha/2}S/\sqrt{R}$. The final step of the QI procedure is to determine whether the c.i. meets the user's half-width requirement, a maximum absolute half-width ϵ' or a maximum relative fraction γ of the magnitude of the final point quantile estimator $\bar{\hat{x}}_p$. If the relevant requirement $w \le \epsilon'$ or $w \le \gamma|\bar{\hat{x}}_p|$ for the precision of the c.i. is satisfied, then the QI procedure terminates, and we return the point quantile estimator $\bar{\hat{x}}_p$ and the c.i. with half-width w. If the precision requirement is not satisfied with R replications, then the QI procedure will increase the number of replications by one. This step can be repeated iteratively until the pre-specified half-width criterion is achieved. This histogram-approximation method can estimate multiple quantiles simultaneously without much extra effort.

6.2.3 *Two-Phase Quantile Estimation*

This section presents the methodologies we will use to construct a c.i. of the quantiles such that they satisfy the absolute or relative half-width requirements, i.e., the quantile estimator $\hat{x}_p$ satisfies the precision requirement $\Pr[x_p \in \hat{x}_p \pm \epsilon'] \ge 1 - \alpha$ or

$$\Pr[x_p \in \hat{x}_p \pm \gamma|\hat{x}_p|] \ge 1 - \alpha, \tag{6.9}$$

where $\epsilon' > 0$ and $0 < \gamma < 1$ are, respectively, the absolute and relative precision sought.

For i.i.d. sequences ($l = 1$) if $N \ge n_p$ (of Eq. (4.6)), then the quantile estimator should satisfy the precision requirement of Eq. (4.5). From Eq. (6.4), asymptotically

$$\mathrm{Var}(\hat{x}_p) \approx \frac{p(1-p)}{N f^2(x_p)}.$$

Therefore, when data are i.i.d.

$$\frac{\text{Var}(\hat{x}_p)}{\text{Var}(\hat{p})} \approx 1/f^2(x_p). \tag{6.10}$$

Thus, if the run length is $N' \geq N/f^2(x_p)$, then $\Pr[x_p \in \hat{x}_p \pm \epsilon''] \geq 1-\alpha$, where ϵ'' has the same numerical value as ϵ (which is unitless) and has the same units as x_p. Moreover, if ϵ' is the desired absolute precision and

$$N' \geq \frac{N}{f^2(x_p)}(\frac{\epsilon''}{\epsilon'})^2 \text{ (or equivalently } \frac{N}{f^2(x_p)}(\frac{\epsilon}{\epsilon'})^2),$$

then (approximately) $\Pr[x_p \in \hat{x}_p \pm \epsilon'] \geq 1 - \alpha$. Note that $0 < \epsilon < 1$, and $\epsilon' > 0$. Furthermore, if γ is the desired relative precision and

$$N' \geq \frac{N}{f^2(x_p)}(\frac{\epsilon''}{\gamma\hat{x}_p})^2 \text{ (or equivalently } \frac{N}{f^2(x_p)}(\frac{\epsilon}{\gamma\hat{x}_p})^2),$$

then (approximately) $\Pr[x_p \in \hat{x}_p \pm \gamma|\hat{x}_p|] \geq 1 - \alpha$.

Theoretically, when N is large, the value of $f(x_p)$ can be approximated by finite forward differences:

$$\widehat{f}(x_p) = \frac{1}{N(\hat{x}_{p+1/N} - \hat{x}_p)} \tag{6.11}$$

because

$$F'(x_p) = \lim_{x \to x_p} \frac{F(x) - F(x_p)}{x - x_p} \approx \frac{F(x_{p+1/N}) - F(x_p)}{x_{p+1/N} - x_p}.$$

Alternatively, the value of $f(x_p)$ can be approximated by finite central differences:

$$\widehat{f}(x_p) = \frac{2}{N(\hat{x}_{p+1/N} - \hat{x}_{p-1/N})} \text{ because } F'(x_p) \approx \frac{F(x_{p+1/N}) - F(x_{p-1/N})}{x_{p+1/N} - x_{p-1/N}}.$$

However, in practice the required value for N is not known for this approximation to be good, so $\widehat{f}(x_p)$ may be very different from $f(x_p)$. The performance of derivative estimation with finite differences using the empirical distribution constructed with the method described in Section 6.2.2 is generally excellent in terms of c.i. coverage and relative precision; see [Chen (2003)]. The value $f(x_p)$ has great influence on the required simulation run length. Because we don't know the value x_p and the pdf $f(\cdot)$ of the underlying process, we use the estimated value $\hat{f}(\hat{x}_p)$. To be conservative, we use the value $\hat{f}(\hat{x}_p)$ such that asymptotically $\Pr[f(x_p) \geq \hat{f}(\hat{x}_p)] \geq 0.6$.

The two-phase quantile estimation algorithm:

(1) The quantity ϵ is the desired proportional $100(1-\alpha_1)\%$ c.i. half-width, γ is the relative precision specified by the user, $100(1-\alpha)\%$ is the desired confidence of coverage with ϵ', and N is the run length.
(2) Use the quantile and histogram estimation algorithm, described in Section 6.2.2, to obtain proportional precision quantile estimates.
(3) Use the finite forward differences Eq. (6.11) to obtain the derivative estimate $\hat{d}_p = \hat{f}(\hat{x}_p)$.
(4) Let $N' = \lceil \frac{N}{\hat{d}_p^2}(\frac{\epsilon}{\gamma \hat{x}_p})^2 \rceil$.
(5) If $N' > N$, increase the run length to N', and go to step 2.
(6) Otherwise, the quantile estimate should already satisfy the relative-precision requirement.
(7) Run R replications and compute the confidence interval of the quantile estimator according to Eq. (6.8).

To improve the precision of the second-phase quantile estimation, we can put more grid points in the grid that contains the quantile estimator $\hat{x}_p$ in the first phase and the surrounding grids before we start the second phase. Of course, the newly set-up grid points need to be based on interpolation. For example, if $g_{i-1} < \hat{x}_p \leq g_i$ and the grid between g_{i-1} and g_i contains η_i observations and approximately $100p_i\%$ of the distribution, then (say) $\kappa + 1$ new grid points can be set up between $\hat{x}_{p-p_i/2}$ and $\hat{x}_{p+p_i/2}$. Let g'_j for $j = 0, 1, \ldots, \kappa$ be the new grid points; then $g'_j = \hat{x}_{p-p_i/2+j(p_i/\kappa)}$. The array contains the number of observations between newly set-up grid points $\eta'_j \approx \eta_i/\kappa$, for $j = 1, 2, \ldots, \kappa - 1$ and $\eta'_\kappa = \eta_i - (\kappa - 1)\eta'_1$.

This method can also be used to determine the required simulation run length to obtain a discrete distributed quantile estimate that satisfies a pre-specified precision. However, the quantile estimate is computed through interpolation so it may not be a valid value of the underlying discrete distribution. If the output data can be read through again, then a valid value can be estimated. In the first phase we obtain lower and upper bounds (Y_a and Y_b) of the quantile. When we read the data again in the second phase, we will count the number of observations that are less than the lower bound, record the values that are between lower and upper bounds, and count the number of observations in each of those values. For example, if there are N observations in total, η_0 observations are less than the lower bound Y_a, Υ values (Y_{a+i}, $i = 1, 2, \ldots, \Upsilon$) are between the lower and upper bound and their corresponding number of observations are η_i (i.e., η_i is the number of observations having the value Y_{a+i}), then the p quantile will be the value Y_j such that $Np \leq \eta_0 + \sum_{i=1}^{j} \eta_i$. Note that the Y_i's correspond to the order statistics of the values of X_i's that are between Y_a and Y_b.

6.2.4 *A Source Code of Quantile Estimation*

```
#define RUNSUPSZ 4000
/* compute the quantile estimators */
int estimatelag(int GRID, int GRID2, nSampSize,
int dist, double beta,double* alpha,double gamma,double* x0,
double* Xp,double* Np,int* gG,double* xSum)
{
int rank;
int i, j, k, l;
int cont = 0;
int runup=1;
double Observ[RUNSUPSZ];
double temp;
long   bufferSize=0;
int    G=*gG;
int    *Np2;
double *Xp2;
double *ObservA;
double *ObservB;

Np2      = (int    *) calloc(G,    sizeof(int)    );
Xp2      = (double *) calloc(G,    sizeof(double));
ObservA  = (double *) calloc(3*nSampSize, sizeof(double));
ObservB  = (double *) calloc(3*nSampSize, sizeof(double));
*xSum    = 0.0;

for (k = 0; k < 3; ++k) {
   l = k + 1;
   for (i = k*nSampSize; i < l*nSampSize; ++i) {
      ObservA[i] = nextrand(dist,1,beta,alpha,gamma,x0);
      *xSum += ObservA[i];
   };
```

```
   for (i = 0; i < RUNSUPSZ; ++i)
      Observ[i] = ObservA[l*i];

   if (checkind(Observ)) {
      cont = -1;       /* break out the while loop */
      k = 3;           /* break out the for loop */
      break;
   }
};

bufferSize = 3*nSampSize;

for (i = 0; i < bufferSize; ++i)
      ObservB[i] = ObservA[i];

qsort(ObservA, (unsigned int) bufferSize,
            sizeof(ObservA[0]), &comparf);

/* set up the grid points */
Xp[0] = ObservA[0];
Xp[G] = ObservA[bufferSize - 1];
Xp2[GRID2+1]=Xp[0];
Np2[GRID2+1]=0;

for (i = 1; i <= GRID; ++i) {
    rank = (int) ceil((1.0/GRID)*i*bufferSize) - 1;
    Xp2[GRID2+1+i] = ObservA[rank];
    Np2[GRID2+1+i] = rank+1;
};

for (i = GRID; i >= 2; --i)
    Np2[GRID2+1+i] = Np2[GRID2+1+i] - Np2[GRID2+i];

/* Set up right tail */
temp = Xp2[GRID+GRID2+1] - Xp2[GRID+GRID2];
for (i = 1; i <= GRID2; ++i) {
    Xp2[GRID+GRID2+1+i] = Xp2[GRID+GRID2+i] + temp;
    Np2[GRID+GRID2+1+i] = 0;
};
```

```
temp = Xp2[GRID2+2] - Xp2[GRID2+1];

for (i = 1; i <= GRID2; ++i) {
    Xp2[GRID2+1-i] = Xp2[GRID2+2-i] - temp;
    Np2[GRID2+1-i] = 0;
};

/* clean up the buffers */
i = 1;
j = 1;
while (j < G) {
    Xp[i] = Xp2[j];
    Np[i] = Np2[j];
    while (++j < G) {
       if (Xp[i] == Xp2[j])
          Np[i] += Np2[j];
       else
          break;
    };
    ++i;
}; /* while j < G */

free(Xp2);
free(Np2);

G = i;
*gG = i;

int p[3];
p[0] = (int) (1.5*nSampSize);
p[1] = 2*nSampSize;
p[2] = 3*nSampSize;

while (++cont) {

   for (i = 1; i < p[0]; ++i)
       ObservB[i] = ObservB[2*i];
```

```
   k = (int) (pow(2,cont) - 1);

   for (l = 0; l < 2; ++l) {

      for (i = p[l]; i < p[l+1]; ++i) {
          ObservB[i] = nextrand(dist,1,beta,alpha,gamma,x0);
          *xSum += ObservB[i];
          bsettable(G, Np, Xp, ObservB[i]);

          for (j = 0; j < k; ++j) {
              temp = nextrand(dist,1,beta,alpha,gamma,x0);
              *xSum += temp;
              bsettable(G, Np, Xp, temp);
          }; /* for j */
      }; /* for i */

      for (i = 0; i < RUNSUPSZ; ++i)
         Observ[i] = ObservB[(l+2)*i];

      if (checkind(Observ)) {
            cont = -1;  /* break out the while loop */
            l = 3;  /* break out the for loop */
            continue;
      };

   }; /* for l */

}; /* while cont */

free (ObservA);
free (ObservB);

return(1);
}



/* function to compare double for qsort */
int comparf(const void *A, const void *B)
```

```
{
double *TA;
double *TB;

   TA = (double *)A;
   TB = (double *)B;

   if (*TA > *TB)
      return (1);

   if (*TA < *TB)
      return (-1);
   else
      return (0);
}



/* procedure to set the Np table in binary order */
void bsettable(int G, double* Np, double* Xp, double target)
/* G is the size of the tables Np and Xp
   Np stores the number of observations between grid points
   Xp stores the value of grid points 0,1,..,G-1
   target is the value to be recorded */
{
int l=1;
int u=G-2;
int b=(1+G-2)/2;
int i=1;

   if (target > Xp[G-2]) {
      ++Np[G-1];
      return;
   } else if (target <= Xp[1]) {
      ++Np[1];
      return;
   }

   while (i) { /* binary search */
```

```
      if (target > Xp[b]) {
         l = b;
         b = (b+u)/2;
         if (l == b)
            ++b;
      } else {
         if (target > Xp[b-1]) {
            ++Np[b];
            i = 0;
            return;
         };
         u = b;
         b = (l+b)/2;
      }; /* if target */
   }; /* while */
}



/* Lagrange interpolation */
double Lagrange(double p, double* Np, double* Xp,
                         int G, double bufferSize)
/* p is the fraction between 0 and 1
   Np stores the number of observations between grid points
   Xp stores the value of grid points 0,1,..,G-1
   G is the size of tables Np and Xp
   bufferSize is the total number of observations */
{
int    i, j, k;
double A[4];
double Y[4];
double P[4];
double temp;
double tempq = p;
double xp;
double size=0;

   for (i = 1;  i < G; ++i) {
       temp = 1.0*Np[i]/bufferSize;
```

```
      if (tempq > temp) {
         size += Np[i];
         tempq -= temp;
         continue;
      } else
         break;
   };

   k = i - 2; /* starting index for Lagrange */

   if (i < 2) {
      xp = Xp[0] + tempq*Xp[1];
      return (xp);
   } else  if (i > G-2) {
      xp = Xp[E-2] + tempq*Xp[E-1];
      return (xp);
   };

   size -= Np[k+1];
   P[0] = 1.0*size / bufferSize; /* up to k=i-2 */
   for (i = 1; i < 4; ++i)
        P[i] = P[i-1] + 1.0*Np[k+i]/bufferSize;

   tempq += P[1];

   xp = 0.0;
   for (i = 0; i < 4; ++i) {
        A[i] = 1.0;
        Y[i] = Xp[k+i];
        for (j = 0; j < 4; ++j) {
            if (i == j) continue;
               A[i] *= (tempq - P[j])/(P[i] - P[j]);
        };
        xp += A[i]*Y[i];
   };

   return (xp);
}
```

6.3 Empirical Experiments

In this section, we present some empirical results obtained from simulations using the two-phase quantile estimation procedure. The purpose of the experiments was not only to test the methods thoroughly, but also to demonstrate the interdependence between the correlation of simulation output sequences and simulation run lengths, and the validity of our methods. We tested the proposed procedure with several i.i.d. and correlated sequences. In these experiments, we use $R = 3$ (see step 7 in the algorithm) independent replications to construct c.i.'s. We estimated four quantile points: 0.25, 0.50, 0.75, and 0.90 for each distribution and used a relative precision of 0.05 for our experiments. We conservatively set the required parameters of determining the simulation run length (i.e., Eq. (4.6)) with $p = 0.5$, $\epsilon = 0.005$, and $\alpha_1 = 0.05$. The confidence level α_2 of the quantile c.i. (i.e., Eq. (6.8)) is set to 0.1. Moreover, the confidence level of the runs-up test of independent is set to 90%.

6.3.1 *Independent Sequences*

We tested two independent sequences:

- Observations are i.i.d. normal with mean 0 and variance 1, denoted as $N(0, 1)$.
- Observations are i.i.d. negative exponential with mean 1, denoted as expon(1).

The summary of our experimental results of the i.i.d. sequences are listed in Tables 6.1 and 6.2. Each design point is based on 100 independent simulation runs. The p row lists the quantile we want to estimate. The *quantile* row lists the true p quantile value. The *cover p* row lists the percentage of the quantile estimates that satisfy Eq. (4.5), i.e., the coverage deviation of the quantile estimator is within the specified value ϵ. The *coverage* row lists the percentage of the c.i.'s that cover the true quantile value. The *avg. rp* row lists the average of the relative precision of the x_p estimators. Here, the relative precision is defined as $rp = |\hat{x}_p - x_p|/|\hat{x}_p|$. The *stdev rp* row lists the standard deviation of the relative precision of the quantile estimators. The *avg. hw* row lists the average of the absolute half-width. The *stdev hw* row lists the standard deviation of the absolute half-width. The *avg. sp* row lists the average of the sample size in each independent replication. The *stdev sp* row lists the standard deviation of

Table 6.1 Coverage of 90% confidence quantile estimators of the $N(0,1)$ distribution

Item	Quantile			
p	0.25	0.45	0.75	0.90
quantile	-0.674189	-0.125381	0.674189	1.28173
cover p	100%	100%	100%	100%
coverage	94%	94%	85%	90%
avg. rp	0.004832	0.010206	0.004693	0.002997
stdev rp	0.003624	0.007715	0.003680	0.002195
avg. hw	0.010011	0.004996	0.008920	0.012771
stdev hw	0.004653	0.002447	0.005363	0.006741
avg. sp	41875	175039	41875	41875
stdev sp	6267	42596	6267	6267

the sample size in each independent replication.

Table 6.1 lists the experimental results of the $N(0,1)$ distribution. For the 0.5 quantile estimates, the parameter under investigation $x_{0.5}$ is 0. Since $\hat{x}_{0.5} \approx 0$, $n' = \lceil \frac{n}{z_p^2}(\frac{\epsilon'}{\gamma \hat{x}_p})^2 \rceil$ will be very large. For example, the sample size in the first phase is 38416 ($1.96^2 \times 0.5 \times 0.5/0.005^2$), the estimator $\hat{x}_{0.5} = -0.000146$, and $z_p = \hat{f}(\hat{x}_{0.5}) = 0.363357$. The required sample size is then $n' = \lceil \frac{38416}{0.363357^2}(\frac{0.005}{0.05 \times 0.000146})^2 \rceil > 1.36 \times 10^{11}$. It will require several days for common desktop computers to obtain one estimator. If users know that the true quantile value $x_p \approx 0$, then absolute precision can be used instead of relative precision. To avoid the required long execution time, we estimated the 0.45 quantile instead. The c.i.'s coverage of 0.75 quantile is 85%, which is less than the specified nominal value of 90%. We believe this is caused by the randomness of the experiment and the half-width being too small. For example, the absolute value of the 0.25 and 0.75 quantile are the same, however, the average half-width of the 0.75 quantile estimates is only 0.008920 compares to 0.010011 of the 0.25 quantile. Furthermore, the average relative precision of the 0.75 quantile estimators is smaller than that of the 0.25 quantile estimators, those 0.75 quantile c.i.'s that do not cover the true quantile value must miss only by a very small amount. All half-widths of the c.i.'s, are less than $\gamma|x_p|$, where $\gamma = 0.05$ and are in general within 30% of $\gamma|x_p|$.

Table 6.2 lists the experimental results of the expon(1) distribution. The sample sizes are the same for all four design points because all quantile estimations have $n' = \lceil \frac{n}{z_p^2}(\frac{\epsilon'}{\gamma \hat{x}_p})^2 \rceil < n$. Since the value $Z_p = f(X_p)$ decreases as X_p increases, sample size n does not increase as quantile value increases when ϵ' is sufficiently small and relative precision is used. On the other hand, if absolute precision is used, then sample sizes will increase

Table 6.2 Coverage of 90% confidence quantile estimators of the *expon*(1) distribution

Item	Quantile			
p	0.25	0.50	0.75	0.90
quantile	0.287682	0.693147	1.38629	2.30258
cover p	100%	100%	100%	100%
coverage	91%	93%	91%	92%
avg. rp	0.004594	0.003241	0.002641	0.002982
stdev rp	0.003691	0.002630	0.002318	0.002186
avg. hw	0.004180	0.007144	0.013503	0.021339
stdev hw	0.001809	0.003672	0.007390	0.011029
avg. sp	41747	41747	41747	41747
stdev sp	6204	6204	6204	6204

as the quantile value increase. In this experiment, with relative precision $\gamma = 0.05$, the quantile estimator obtained with the first-phase sample size should satisfy both Eq. (4.5) and Eq. (6.9). Again, all quantile estimators satisfy the precision requirement of Eq. (4.5), and the c.i. coverage of these design points are above the specified 90% confidence level. Furthermore, all half-widths are less than $\gamma|x_p|$. Since the true p quantile value increases as p increases, the average half-width increases as p increases. Because we set the confidence level of the runs-up test of independence to be 90%, independent sequences will not pass the runs-up about 10% of the times. Consequently, the first-phase sample size for independent sequences will be around $38416 \times 1.1 = 42258$.

6.3.2 *Correlated Sequences*

We tested four correlated sequences:

- Steady-state of the *first-order moving average* process, generated by the recurrence relation

$$X_i = \mu + \epsilon_i + \theta\epsilon_{i-1} \quad \text{for} \quad i = 1, 2, \ldots,$$

where ϵ_i is i.i.d. $N(0,1)$ and $0 < \theta < 1$. This process is denoted as MA1(θ). μ is set to 0 in our experiments. It can be shown that X has an asymptotic $N(0, 1+\theta^2)$ distribution.

- Steady-state of the *first-order autoregressive* process, generated by the recurrence relation

$$X_i = \mu + \varphi(X_{i-1} - \mu) + \epsilon_i \text{ for } i = 1, 2, \ldots,$$

where ϵ_i is i.i.d. $N(0,1)$, and

$$E(\epsilon_i) = 0, \quad E(\epsilon_i \epsilon_j) = \begin{cases} \sigma^2 \text{ if } i = j \ , \\ 0 \quad \text{otherwise} \end{cases}$$

$$0 < \varphi < 1.$$

This process is denoted as AR1(φ). μ is set to 0 in our experiments. It can be shown that X has an asymptotic $N(0, \frac{1}{1-\varphi^2})$ distribution.

- Steady-state of the M/M/1 delay-in-queue process with the arrival rate (λ) and the service rate ($\nu = 1$). This process is denoted as MM1(λ). Let W_i denote the waiting time of the ith customer and $\rho = \lambda/\nu$ be the traffic intensity. Then, if $\rho < 1$, the theoretical steady-state distribution of this M/M/1 queuing process is $F(x) = P(W_i \leq x) \to 1 - \rho e^{-(\nu-\lambda)x}$ as $i \to \infty$ for all $x \geq 0$. Let $\{A_n\}$ denote the interarrival-time i.i.d. sequence and $\{S_n\}$ denote the service-time i.i.d. sequence. Then the waiting-time sequence $\{W_n\}$ is defined by

$$W_{n+1} = \ (W_n + S_n - A_{n+1})^+ \ \text{ for } \quad n \geq 1$$

where $w^+ = \max(w, 0)$.

- Steady-state of the M/M/s delay-in-queue process with the arrival rate (λ) and the service rate (ν). This process is denoted as MMS(λ). We set $s = 2$, $\lambda = 3$, and $\nu = 2$. The traffic intensity of this process is $\rho = \frac{\lambda}{s\nu} = 0.75$.

We tested the MA1 model with $\theta = 0.75$, the AR1 model with $\varphi = 0.75$, and the M/M/1 and M/M/2 models with the traffic intensity $\rho = 0.75$. In order to eliminate the initial bias, ϵ_0 and w_0 are set to a random variate drawn from the steady-state distribution. Because the true 0.5 quantile value for the tested MA1 and AR1 processes is 0, we estimated the 0.45 quantile to avoid an extremely large sample size.

The summary of our experimental results for MA1(0.75) and AR1(0.75) are listed in Tables 6.3 and 6.4. All quantile estimators satisfy the precision requirement of Eq. (4.5). Most of the c.i. coverage of these design points, except the coverage of the 0.75 quantile of the AR1(0.75) process, are around the specified 90% confidence level. We believe this is caused by the half-width being too small. For these two processes, $n' = \lceil \frac{n}{z_p^2} (\frac{\epsilon'}{\gamma \hat{x}_p})^2 \rceil > n$ only when estimating the 0.45 quantile. The sample sizes are larger than the independent cases because the lag l for the QI sequence that appears to be independent is larger.

Table 6.3 Coverage of 90% confidence quantile estimators of the $MA1(0.75)$ process

Item	Quantile			
p	0.25	0.45	0.75	0.90
quantile	-0.842737	-0.156726	0.842737	1.60216
cover p	100%	100%	100%	100%
coverage	91%	93%	90%	88%
avg. rp	0.004897	0.010006	0.004437	0.002558
stdev rp	0.003227	0.007751	0.003252	0.001905
avg. hw	0.011336	0.005633	0.010974	0.013519
stdev hw	0.005510	0.002892	0.005519	0.006688
avg. sp	80805	326563	80805	80805
stdev sp	6977	42356	6977	6977

Table 6.4 Coverage of 90% confidence quantile estimators of the $AR1(0.75)$ process

Item	Quantile			
p	0.25	0.45	0.75	0.90
quantile	-1.01928	-0.189558	1.01928	1.93779
cover p	100%	100%	100%	100%
coverage	91%	92%	82%	91%
avg. rp	0.003340	0.009194	0.003583	0.001861
stdev rp	0.002279	0.007469	0.002411	0.001473
avg. hw	0.012107	0.005560	0.010085	0.012185
stdev hw	0.005923	0.002606	0.005668	0.006173
avg. sp	348067	1424348	348067	348067
stdev sp	54251	344237	54251	54251

If $\rho < 1$, the waiting-time distribution function of a stationary M/M/1 delay in queue is discontinuous at $F(x) \to 1 - \rho$, (i.e. $x = 0$); thus, the quantiles for M/M/1 delay-in-queue are applicable only when the estimated quantiles are larger than or equal to $1 - \rho$. Therefore, it is useful to know whether a desired quantile is attainable before conducting an informative experiment.

The summary of our experimental results of the M/M/1 delay-in-queue process is summarized in Table 6.5. We experienced some problems when estimating the 0.25 quantile of the M/M/1 queuing process with $\rho = 0.75$, because the distribution is not continuous at the true quantile value 0. Thus, the derivative does not exist at 0.25 quantile. Because the distribution function has a jump at this quantile point, the procedure often obtains ∞ as an estimate of the derivative since $\hat{x}_p = \hat{x}_{p+1/N}$ in this case. Therefore, we estimate the 0.30 quantile instead of the 0.25 quantile. However, the procedure can return the quantile estimate obtained in the first

Table 6.5 Coverage of 90% confidence quantile estimators of the $M/M/1$ delay-in-queue process with $\rho = 0.75$

Item	Quantile			
p	0.30	0.50	0.75	0.90
quantile	0.275972	1.62186	4.39445	8.05961
cover p	100%	100%	100%	100%
coverage	90%	91%	92%	90%
avg. rp	0.005469	0.004432	0.003583	0.003919
stdev rp	0.004258	0.003228	0.002532	0.002947
avg. hw	0.005576	0.021964	0.051527	0.105334
stdev hw	0.003185	0.012968	0.029221	0.058994
avg. sp	5491879	1363070	1363070	1363070
stdev sp	1541480	297519	297519	297519

phase. Users should then investigate if the distribution is continuous at this particular quantile. Again, all quantile estimators satisfy the precision requirement of Eq. (4.5), and c.i. coverages are above or close to the specified 90%. The average c.i. half-width of the 0.90 quantiles of the M/M/1 delay in queue is much larger than the other quantiles since the quantile under estimation has a larger value.

The summary of our experimental results of the M/M/2 delay-in-queue process is listed in Table 6.6. If $\rho < 1$, the theoretical steady-state distribution of this M/M/2 queuing process is $F(x) \to 1 - 9e^{-x}/14$, where $x \geq 0$. Therefore, for this M/M/2 process quantiles less than 5/14 are not attainable, we estimated 0.40 quantile instead. All estimators satisfy the probability coverage requirements. Moreover, the percentages of the c.i.'s that cover the true quantiles are close to the specified nominal value of 90%. The sample size determined by the QI procedure is roughly the same for the waiting-time of M/M/2 and the M/M/1 delay in queue with the same traffic intensity of $\rho = 0.75$. However, the c.i. coverage of M/M/2 delay in queue is not as good as M/M/1. Again, we believe this is caused by the half-width being too small.

6.3.3 *A Practical Application*

In this section, we propose a new approach to estimate the critical constant for [Rinott (1978)] procedure.

For $l = 1, 2, \ldots, k$, let z_{i_l} and χ_{i_l} denote the variables having the standard normal distribution and the χ^2 distribution with $n_0 - 1$ d.f., respec-

Table 6.6 Coverage of 90% confidence quantile estimators of the $M/M/2$ delay-in-queue process with $\rho = 0.75$

Precision	Traffic Intensity ρ			
	0.75			
p	0.40	0.50	0.75	0.90
quantile	0.068993	0.251314	0.944462	1.86075
cover p	100%	100%	100%	100%
coverage	89%	90%	88%	84%
avg. rp	0.006573	0.002424	0.004137	0.004463
stdev rp	0.005017	0.001859	0.002950	0.003152
avg. hw	0.001627	0.002265	0.012300	0.024968
stdev hw	0.000828	0.001122	0.006728	0.014283
avg. sp	7925708	1326701	1326701	1326701
stdev sp	2921767	285481	285481	285481

tively. Let

$$T_{i_l} = \frac{z_{i_l}}{\sqrt{\chi_{i_l}/(n_0 - 1)}}.$$

The variables T_{i_l}'s are i.i.d. t-distributed with $n_0 - 1$ d.f. The selection procedure of [Dudewicz and Dalal (1975)] is derived from the equation

$$\mathrm{P(CS)} \geq \mathrm{P}[T_{i_1} < T_{i_l} + h_1, l = 2, 3, \ldots, k] = \int_{-\infty}^{\infty} F^{k-1}(t + h_1) f(t) dt,$$

where f and F are the pdf and cdf of the t distribution with $n_0 - 1$ d.f., respectively. This equality holds exactly under the LFC.

We follow the steps of Proposition 3 of [Rinott (1978)] to show that under the LFC

$$\begin{aligned}
&\mathrm{P(CS)} \\
&= \mathrm{P}[T_{i_1} < T_{i_l} + h_r, l = 2, 3, \ldots, k] \\
&= \mathrm{P}[\frac{T_{i_1} - T_{i_l}}{\sqrt{(n_0 - 1)(1/\chi_{i_1} + 1/\chi_{i_l})}} < \frac{h_r}{\sqrt{(n_0 - 1)(1/\chi_{i_1} + 1/\chi_{i_l})}}, l = 2, 3, \ldots, k] \\
&\geq \Phi(\frac{h_r}{\sqrt{(n_0 - 1)(1/\chi_{i_1} + 1/\chi_{i_l})}}, l = 2, 3, \ldots, k) \\
&= \mathrm{P}[z_{i_l} \sqrt{(n_0 - 1)(1/\chi_{i_1} + 1/\chi_{i_l})} < h_r, l = 2, 3, \ldots, k].
\end{aligned}$$

The inequality follows Slepian's inequality because variables

$$Z'_{i_l} = \frac{T_{i_1} - T_{i_l}}{\sqrt{(n_0 - 1)(1/\chi_{i_1} + 1/\chi_{i_l})}} \sim N(0, 1)$$

and Z'_{i_l}'s are correlated. The last equality follows since z_{i_l} for $l = 2, 3, \ldots, k$ are i.i.d. $N(0, 1)$ variables. Consequently, the value of h_1 such that

$\int_{-\infty}^{\infty} F^{k-1}(t + h_1)f(t)dt = P^*$ is no larger than the value of h_r such that $\mathrm{P}[z_{i_l}\sqrt{(n_0 - 1)(1/\chi_{i_1} + 1/\chi_{i_l})} < h_r, l = 2, 3, \ldots, k] = P^*$. That is, $h_1 \leq h_r$.

The h_r values can be reduced for [Rinott (1978)] procedure by taking into account the correlation between Z'_{i_l}'s. Let $\xi_{i_l} \sim N(0, 0.5)$ for $l = 1, 2, \ldots, k$ and let $Z''_{i_l} = \xi_{i_l} - \xi_{i_1}$ for $l = 2, 3, \ldots, k$. Note that $Z''_{i_l} \sim N(0, 1)$ and correlates with the correlation coefficient $1/2$, i.e.,

$$\mathrm{Cov}(Z''_{i_a}, Z''_{i_b}) = (\frac{0.5}{0.5} + 1)^{-1/2}(\frac{0.5}{0.5} + 1)^{-1/2}, \ a \neq b.$$

Let $\Omega = \max_{l=2}^{k} Z''_{i_l}\sqrt{(n_0 - 1)(1/\chi_{i_1} + 1/\chi_{i_l})}$. The value of h_r is then computed such that

$$\mathrm{P}[\Omega < h_r] = P^*. \tag{6.12}$$

The estimated h_r values are available in [Chen (2011)]. These h_r values are smaller than the corresponding values listed in [Wilcox (1984)] and are slightly larger than the corresponding h_1 values for the procedure of [Dudewicz and Dalal (1975)].

Without the knowledge of $\sigma^2_{i_l}$'s, letting $\xi_{i_l} \sim N(0, 0.5)$ for $l = 1, 2, \ldots, k$ is a suitable alternative. If $\xi_{i_1} \sim N(0, 0)$ and $\xi_{i_l} \sim N(0, 1)$ for $l = 2, 3, \ldots, k$, then Z''_{i_l}'s are independent and the corresponding h_r values are the same as those listed in [Wilcox (1984)]. On the other hand, if $\xi_{i_1} \sim N(0, 1)$ and $\xi_{i_l} \sim N(0, 0)$ for $l = 2, 3, \ldots, k$, then Z''_{i_l}'s degenerate to a single variable and the corresponding h_r values will be smaller than the true values.

6.4 Summary

We have presented an algorithm for estimating the histogram of a stationary process. Some histogram estimates require more observations than others before the asymptotic approximation becomes valid. The proposed quasi-independent algorithm works well in determining the required simulation run length for the asymptotic approximation to become valid. The QI procedure estimates the required sample size based entirely on data and does not require any user intervention. Moreover, the QI procedure processes each observation only once and does not require storing the entire output sequence. Since the procedure stops when the QI subsequence appears to be independent, the procedure obtains high precision and small half-width with long simulation run length by default.

The histogram-approximation algorithm computes quantiles only at certain grid points and generates an empirical distribution (histogram) of the output sequence, which can provide valuable insights of the underlying stochastic process. Because the QI procedure does not need to read the output sequence repeatedly, it is an online algorithm and the storage requirement is minimal. The main advantage of the approach is that by using a straightforward runs test to determine the simulation run length and using natural estimators to construct the c.i., we can apply classical statistical techniques directly and do not require more advanced statistical theory, thus making it easy to understand and simple to implement.

This algorithm has been implemented in simulation software packages, e.g., the BigHouse Simulator [Meisner et al. (2012)]. Because a histogram is constructed as an empirical distribution of the underlying process, it is possible to estimate multiple quantiles simultaneously,and estimate other characteristics of the distribution, such as a proportion, or derivative [Chen (2003)], under the same framework. Preliminary experimental results indicate that the natural estimators obtained based on the empirical distribution are fairly accurate.

Chapter 7

Metamodels

This chapter reviews metamodels, investigates the accuracy of the fitted quantile curves (response surface) and how to construct non-functional-form metamodels with a set of histograms at certain design points.

7.1 Introduction

In many cases, users are interested in the system responses under different design points (input combinations, scenarios). Thus, a series of design points need to be evaluated, i.e., we need to construct a metamodel (or a response surface). For example, we are interested in the mean waiting time (or system time) of queuing systems with different traffic intensities. The metamodel can be "used as a proxy for the full-blown simulation itself in order to get at least a rough idea of what would happen for a large number of design points".

The purpose of constructing metamodels is to estimate or approximate the response surface. We could then use the metamodel to learn how the response surface would behave over various input-parameter combinations, such as output sensitivity. This approach is "helpful when the simulation is very large and costly, precluding exploration of all but a few input-parameter combinations". Note that the metamodel is designed to provide the overall tendency of performance measures rather than accurate estimates at all input-parameter combinations. These metamodels can also be used for visualization. Graphical representations of metamodels are useful for providing a simple and easy form to communicate the input-output relationship.

Traditionally, a metamodel is specified to be a standard regression model and is described by a formula obtained through regression on several chosen design points, see, e.g., [Turner et al. (2013)]. However, we usually do not know what functional form to specify for the regression terms. Furthermore, the value at design points obtained via the metamodel (formula) may be different from the original observed value at the design point. On the other hand, Kriging [van Beers and Kleijnen (2008)] is a metamodeling methodology (originally developed for deterministic simulation) that estimates the value at non-design points by interpolation. Since no regression is performed to obtain a formula, for selected design points (input combinations), the Kriging metamodel simply returns the observed values. In non-deterministic simulation, the observed value at design points may be the average of several replications. For a non-design point, the interpolated value is a weighted average of the values observed at all design points. [Jones et al. (1998)] detail the difference between regression-based metamodels and interpolation-based metamodels.

Kriging requires a correlation function (i.e., the spatial dependence) to compute the weights, which depend on the distances between the input combination to be estimated and the existing input combinations already observed. Kriging assumes that the closer the input combinations are, the stronger positively correlated the responses are. The choice of correlation function should be motivated by the underlying system we want to model. However, the (true) correlation function is unknown and both its type and parameter values must be estimated. Furthermore, some of the weights may be negative. This correlation function is similar to the kernel function used to smooth estimates when estimating density. See [van Beers and Kleijnen (2008)] for an overview of Kriging.

Design of experiments can be used during the metamodel construction to improve the construction process and quality of the metamodel. An important issue in metamodeling is how to select the design points and estimate their responses to which the metamodel is fitted. We investigate the issue of finding both the number and the placement of design points to achieve the required precision. To account for the stochastic nature of the output, two additional metamodels are constructed: lower and upper bounds. For non-design points, the interpolated value (at each dimension) is a linear interpolation of two bounding design points, not a weighted average of all design points. Hence, the value at non-design points can be estimated inexpensively. This approach is a special case of the correlation function LIN in the Matlab Kriging toolbox developed by [Lophaven et al. (2002)].

7.2 Constructing Metamodels of Quantiles

A simulation model can be thought of as a *function* that turns input parameters into output performance measures. For example, if we use simulation to estimate x_p (the p quantile) of the M/M/1 queuing process with certain traffic intensity $0 < \rho < 1$, we could in principle write

$$x_p = F(p, \rho)$$

for some function F that is stochastic and unknown and we use simulation to evaluate F for numerical input values of p and ρ. We are interested in the p quantile x_p for any point in the two-dimensional region defined by the proportion p and the traffic intensity ρ. Both p and ρ are continuous, it is impossible to evaluate all combinations of p and ρ numerically. In some other cases, it maybe possible to evaluate all input combination numerically; but most likely is not practical to do so. Instead, we evaluate the function at certain input combinations and fit the results to some curves. The response of other input combinations is then approximated via the fitted curves (surfaces).

It is nice to have a single formula to represent the responses of a system, but the formula is generally unknown and is likely complex. Fitting the grid points to a formula also introduces error into the estimates. In the approach, instead of obtaining a formula or a regression model, we treat the collection of grid points themselves and the set of fitted curves as metamodels. Consequently, the metamodel is not described by a single (simple) formula. We call the metamodel of this kind the non-functional-form metamodel. For example, there is no closed form of the cdf of the standard normal distribution and the quantiles are customarily listed in tables instead of by a regression model. Without the help of a calculator, the quantile value can not be easily computed from the formula.

A metamodel will have greater accuracy and precision when constructed with a larger set of input-parameter combinations. On the other hand, a larger set of input-parameter combinations requires more simulation efforts. Furthermore, the required number of input-parameter combinations of constructing metamodels to achieve the pre-specified accuracy and precision depends on the underlying systems. We can sequentially determine the required number of input-parameter combinations by observing the fitted curves. Even though both p and ρ are input parameters, only ρ is designated as a design point during the metamodel construction; since $\rho = \lambda/\nu$ can be controlled by setting the values of arrival rate λ or server rate ν. The value at non-design points will be estimated via interpolation. Furthermore, the randomness of the response at design points is described by confidence intervals. Consequently, three response surfaces will be constructed: the lower bound, the expected value, and the upper bound. It is likely that the form and complexity of the metamodel vary substantially for different systems. Thus, the metamodeling procedure needs to include a scheme for model selection: given the simulation data, obtain a metamodel that is of the least complexity but adequate to characterize the underlying response surface.

Assuming the range of interest of the (controllable) parameter ρ is $[\rho_L, \rho_U]$, we initially simulate the system with five evenly spaced design points, e.g., ρ (say $\rho_1 = \rho_L < \rho_2 < \cdots < \rho_5 = \rho_U$). Let F_i be the simulated (estimated) response of input combination (p, ρ_i) for some selected p. We recommend the selected p should include at least five evenly spaced points between the range of interest. We fit curves C_1 and C_2, respectively, to the three $((p, \rho_1), (p, \rho_3), (p, \rho_5))$ and five evenly spaced input-parameter combinations and their corresponding responses F_i. We then use curve C_1 to estimate the response of (p, ρ_2) and (p, ρ_4) and let $\hat{F}_2$ and $\hat{F}_4$ be the estimated responses from C_1. If the relative precision

$$|\hat{F}_i - F_i|/F_i < 4\gamma \tag{7.1}$$

is satisfied for all i, then the procedure returns the collection of all the histograms as the non-functional-form metamodel; otherwise design points $\rho_{j_1} = (\rho_{i-1} + \rho_i)/2$ and $\rho_{j_2} = (\rho_i + \rho_{i+1})/2$ will be added for each i that the precision requirement is not satisfied. Note that the value γ is the intended relative precision of quantile estimates, i.e.,

$$\frac{|\bar{\hat{x}}_p - x_p|}{x_p} \leq \gamma,$$

where $\bar{\hat{x}}_p$ is the final point estimator of x_p, see Eq. (6.7). Using the value 4γ in Eq. (7.1) is somewhat arbitrary. The rationale are that the distance between the design points ρ_{i-1} and ρ_i is half of the distance (between ρ_{i-1} and ρ_{i+1}) used to estimate $\hat{F}_i$ and the angle of the line between design points is also reduced by no less than half. The step of adding more design points will be performed repeatedly until the specified precision on the selected quantile estimators is achieved. Thus, the final number of design points is not known until the simulation is complete. The preliminary experimental results indicate that the required design points to achieve the specified precision is not excessive.

Table 7.1 Chosen design points

Procedure	Traffic Intensity									
LI	0.1	0.3	0.4	0.5	0.6	0.7	0.75	0.8	0.85	0.9
V&K	0.1	0.3	0.5	0.7	0.8	0.85	0.875	0.8875	0.89375	0.9

7.3 Constructing Quantile Confidence Interval

With a given design point (e.g., traffic intensity ρ), we use the Histogram Approximation (HA) procedure discussed in Chapter 6 to construct histograms and estimate quantiles. We construct the curve of quantiles, the curves of the lower and upper confidence limits of quantiles at the design points. The quantile point estimator and c.i. at non-design points can then be obtained via linear interpolation. Let x_p and y_p, respectively, denote the p quantile estimators of the time in system of the M/M/1 queuing process with traffic intensity ρ_1 and ρ_2. Then the p quantile estimator with traffic intensity $\rho_1 < \rho < \rho_2$ is estimated by linear interpolation, i.e.,

$$z_p = x_p + \frac{\rho - \rho_1}{\rho_2 - \rho_1}(y_p - x_p).$$

The confidence limits of z_p are computed in the same manner. [Chen and Li (2014)] discuss how to apply this procedure in higher dimensions.

7.4 Empirical experiments

In this section, we present some empirical results obtained from simulations using non-functional-form metamodels to estimate quantiles.

7.4.1 *Choosing the Design Points*

In order to compare our customized design with other approaches, we perform a similar experiment as in [van Beers and Kleijnen (2008)]. We want to construct a metamodel of the waiting-time of the M/M/1 queuing system for traffic intensity $0.1 \le \rho \le 0.9$. We begin with five evenly spaced traffic intensity (i.e., pilot runs) and sequentially increase the number of design points until the number of design points reach 10, the same number as in [van Beers and Kleijnen (2008)] for comparison. In the experiment, each response is the true response instead of the simulated response.

Let $\gamma_i = |\hat{F}_i - F_i|/F_i$. To adapt to this stopping rule, we add the design point $\rho_{j_1} = (\rho_{j-1} + \rho_j)/2$ or $\rho_{j_2} = (\rho_j + \rho_{j+1})/2$ when $\gamma_j = \max \gamma_i$. Table

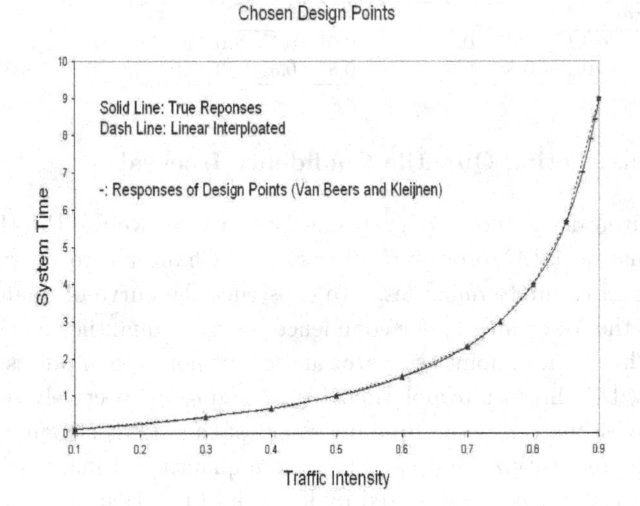

Fig. 7.1 Plots of the linear interpolated versus true M/M/1 system time

7.1 lists the final 10 design points. The LI and V&K rows, respectively, are for the linear interpolated and the procedure of [van Beers and Kleijnen (2008)]. The LI approach places design point more evenly because the strategy basically aims to minimize the maximum of γ_i and selects relatively few design points in the area that the responses change linearly as the input changes. On the other hand, the approach of V&K aims to minimize the integrated mean square error and is computationally expensive. Consequently, they place all the additional design points between traffic intensities 0.7 and 0.9, where the responses have larger values and larger variances. Figure 7.1 displays the true responses and the linear-interpolated responses.

7.4.2 *Estimating Quantiles of Moving-Average and Autoregressive Processes via Non-functional-form Metamodels*

In this experiment, we present some empirical results of estimating quantiles of the *first-order moving-average* MA1(θ) and the *first-order autoregressive* AR1(φ) processes obtained from non-functional-form metamodels.

The MA1(θ) process is generated by the sequence

$$X_i = \mu + \epsilon_i + \theta\epsilon_{i-1} \quad \text{for} \quad i = 1, 2, \ldots.$$

The AR1(φ) process is generated by the recurrence relation

$$X_i = \mu + \varphi(X_{i-1} - \mu) + \epsilon_i \text{ for } i = 1, 2, \ldots,$$

where X_0 is specified as a random variate drawn from the steady-state distribution. In both process we set μ to 2 and ϵ_i to be i.i.d. $N(0, 1)$.

The values of the required parameters of determining the simulation run length (i.e., Eq. (4.6)) are $p = 0.5$, $\epsilon = 0.01$, and $\alpha_1 = 0.05$. The confidence level of the test of independence is set to 0.95^2. In these experiments, the point quantile estimate (i.e., $\bar{\hat{x}}_p$) at design points is the average of $R = 3$ independent replications. We use relative precision (i.e., Eq. (7.1) with $\gamma = 5\%$) to determine the number of design points.

We are interested in quantiles of these processes with correlation coefficient between 0.7 and 0.95. Since at least five design points should be used, the initial difference between design points is 0.0625 (i.e., (0.95-0.7)/4). We construct histograms at the following design points (i.e., the correlation coefficients φ and θ): 0.7, 0.7625, 0.825, 0.8875, and 0.95 and estimate 0.85, 0.9, and 0.95 quantiles at the following correlation coefficients: 0.7, 0.75, 0.8, 0.85, 0.9, and 0.95. No additional design points are included because the relative precision of quantile estimates from the histogram at correlation coefficient 0.7625 (0.8875) and from linear interpolation with histograms at correlation coefficients 0.7 and 0.825 (0.825 and 0.95) are within the specified relative precision $4\gamma = 20\%$, see Section 7.2. The quantile estimates at 0.7 and 0.95 correlation coefficients are obtained directly from the histograms. The quantile estimates at other correlation coefficients are obtained through linear interpolation.

The left-hand-side graphs in Figure 7.2 show the quantile-estimation results for MA1. They are relative deviation plots with the p quantile being 0.85, 0.90, and 0.95. For these graphs, the x-axis represents the correlation coefficient θ, the y-axis represents the relative deviation, i.e.,

$$\frac{\bar{\hat{x}}_p - x_p}{x_p} \times 100\%,$$

and every point (p, x_p) in the graph represents the observed relative deviation at certain θ from one of the 100 independent simulation runs. The right-hand-side graphs in Figure 7.2 show the quantile-estimation results for AR1.

For the MA1 process, the p quantile changes approximately linearly as θ changes, hence, the observed relative deviations are roughly the same at all θ. The relative precisions are all within 1%. For the AR1 process, the p quantile changes non-linearly as φ changes, likely concave upward. The accuracy of quantile estimates obtained from the histograms at the design points (i.e., at $\varphi = 0.7$ and 0.95) is high indicating the performance of the HA procedure is good. The accuracy of the estimates degraded when estimating quantile of input-parameter combinations that are not design points. This is not unexpected because using interpolation introduces errors. Nevertheless, the relative precisions are well within the specified $\gamma = 5\%$.

7.4.3 *Estimating Quantiles of Queuing Systems via Non-functional-form Metamodels*

We estimate the steady-state waiting time in system quantile of the M/M/1 queuing systems. The values of the required parameters of determining the simulation run length are as set in the previous experiments.

We construct histograms at 0.7, 0.7625, 0.825, 0.8563 ($\approx$ (0.825 + 0.8875)/2), 0.8875, 0.9188 ($\approx$ (0.8875 + 0.95)/2), and 0.95 traffic intensities (i.e., design points) and estimate 0.85, 0.9, and 0.95 quantiles at 0.7, 0.75, 0.8, 0.85, 0.9, and 0.95 traffic intensities. Traffic intensities 0.8563 and 0.9188 are included in the design points because the relative precision of quantile estimates from the histogram at traffic intensity of 0.8875 and from linear interpolation with histograms at traffic intensities 0.825 and 0.925 is larger then the specified relative precision $4\gamma = 20\%$, see Section 7.2.

Two types of plots were made to display graphically the 100 realizations of each quantile estimator: 1) relative deviation plots; and 2) absolute deviation plots, in which quantile estimates are plotted around their true values. The left-hand-side and right-hand-side graphs on Figure 7.3 show the relative plots and the absolute plots, respectively. In the absolute deviation plots, the solid curve represents a piecewise linear version of the true quantile curve across various traffic intensities and the quantile estimates are plotted as a point. Note that a piecewise linear version of quantile estimates can provide information regarding the general tendency of quantiles and a rough idea of quantile sensitivity as the traffic intensity changes. For the M/M/1 queuing process, the quantile estimates obtained through linear interpolation are biased high, which can be explained by the curves in the

Table 7.2 Estimated 90% confidence interval half width and coverage for M/M/1

	0.85 Quantile		0.90 Quantile		0.95 Quantile	
ρ	Avg HW	Coverage	Avg Hw	Coverage	Avg HW	Coverage
0.70	0.115	0.86	0.153	0.89	0.235	0.88
0.75	0.149	0.89	0.188	0.88	0.291	0.95
0.80	0.185	0.29	0.244	0.48	0.371	0.58
0.85	0.241	0.90	0.319	0.89	0.481	0.88
0.90	0.347	0.21	0.463	0.30	0.747	0.42
0.95	0.716	0.90	0.943	0.90	1.538	0.92

right-hand-side graphs on Figure 7.5. Given a fixed p, the p quantile vs. ρ curve is concave upward, i.e., given a fixed p $F(p, \rho)$ is a convex function. As the value of ρ deviates further away from the design points, e.g., $\rho = 0.80$ and 0.90, the accuracy of the estimates gets worse. Note that the (absolute or relative) precision can be improved by increasing the number of design points. The relative precision of all estimates from the histogram approximated non-functional-form metamodel are well within the specified $\gamma = 5\%$.

A c.i. is constructed for each quantile with $R = 3$ independent quantile estimators. Table 7.2 shows the results. The column labeled "Avg HW" is the average of the 90% c.i. half width calculated from the 100 realizations of the quantile estimator. The "Coverage" column lists the proportion of these 100 c.i.'s that cover the true quantile. The c.i. coverage at the design points (i.e., $\rho = 0.70$ and 0.95) are around the nominal value of 0.90, which indicates the estimated c.i. half widths are also accurate. However, the c.i. coverages of ρ that are further away from the design points (e.g., $\rho = 0.80$ and 0.90) are less than the nominal value. This is expected since the point estimators are bias high; while the half widths are about the right length. We don't think this is a major drawback of the procedure. The main purpose of the metamodel is to gain the overall tendency of quantiles of the system under study and is not to obtain accurate quantile c.i.'s throughout the entire traffic-intensity range of interest. Note that accurate quantile c.i.'s can be obtained by designating the traffic intensity of interest ρ as a design point. Furthermore, with the help of the quantile plots, we are able to predict approximately the accuracy of the interpolated quantile estimates.

If the c.i. coverage at non-design points is a concern, a conservative adjustment can be used to increase the coverage of the c.i. estimated via interpolation. Let $\gamma_i = |\hat{F}_i - F_i|/F_i$. When using interpolation with curves

C_{i-1} and C_i or C_i and C_{i+1} to estimate c.i., we set

$$\text{c.i.} = \begin{cases} (\bar{\bar{x}}_p - w - \gamma_i \bar{\bar{x}}_p, \bar{\bar{x}}_p + w) & \hat{F}_i > F_i \\ (\bar{\bar{x}}_p - w, \bar{\bar{x}}_p + w + \gamma_i \bar{\bar{x}}_p) & \hat{F}_i < F_i. \end{cases}$$

While this adjusted c.i. will increase the coverage to be greater than the nominal value, the range of the c.i. will be much larger.

7.5 Summary

Non-functional-form metamodels of providing overall tendency of quantiles can be constructed with a set of carefully selected histograms. The metamodel can be used as a proxy for the full-blown simulation itself in order to get at least a rough idea of what would happen for a large number of input-parameter combinations. Estimates obtained via linear interpolation of grid points of metamodels are as good as those obtained via other sophisticated methods when the structure of the underlying systems is unknown.

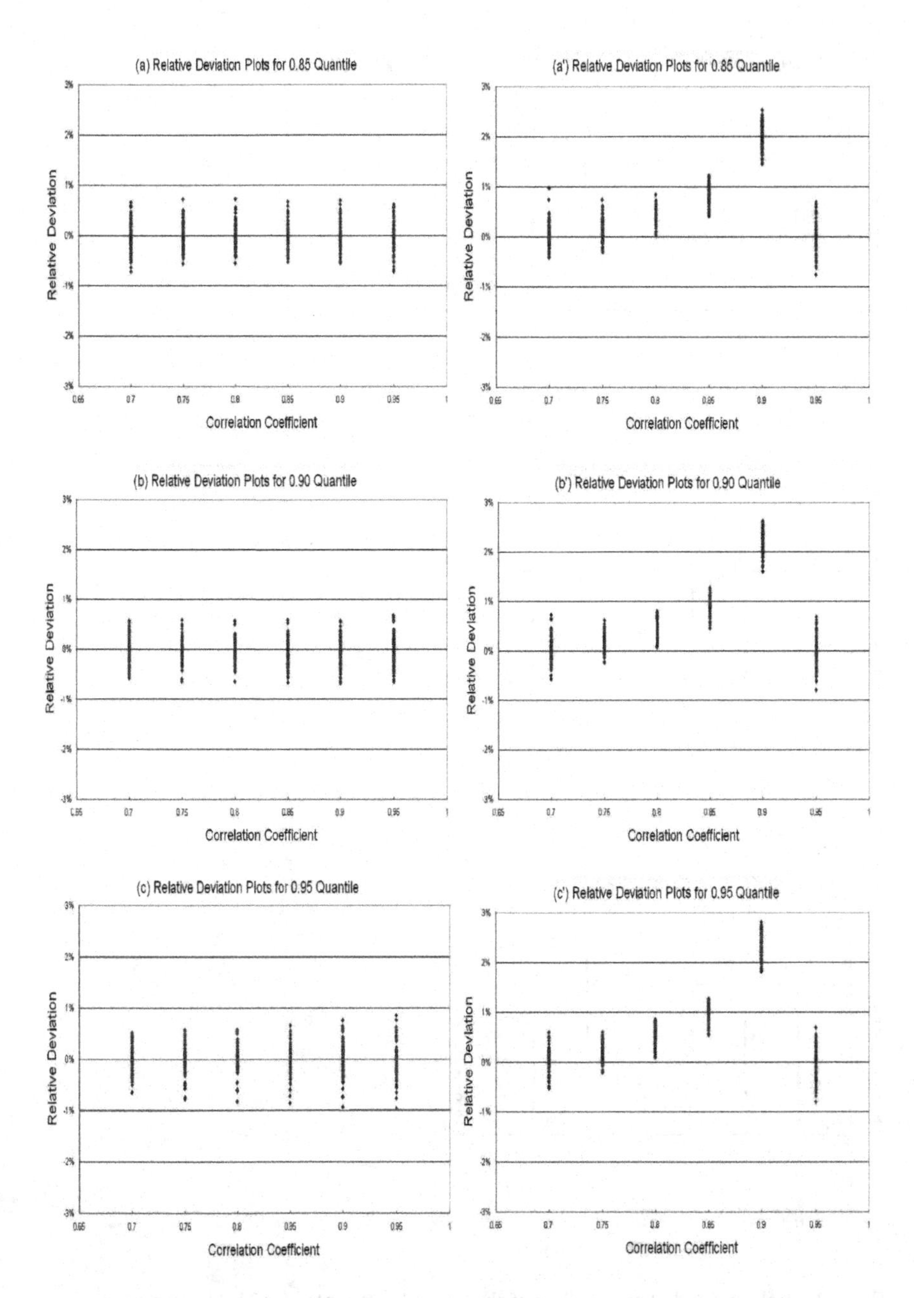

Fig. 7.2 Plots of the quantile estimates for MA1 and AR1 via metamodels

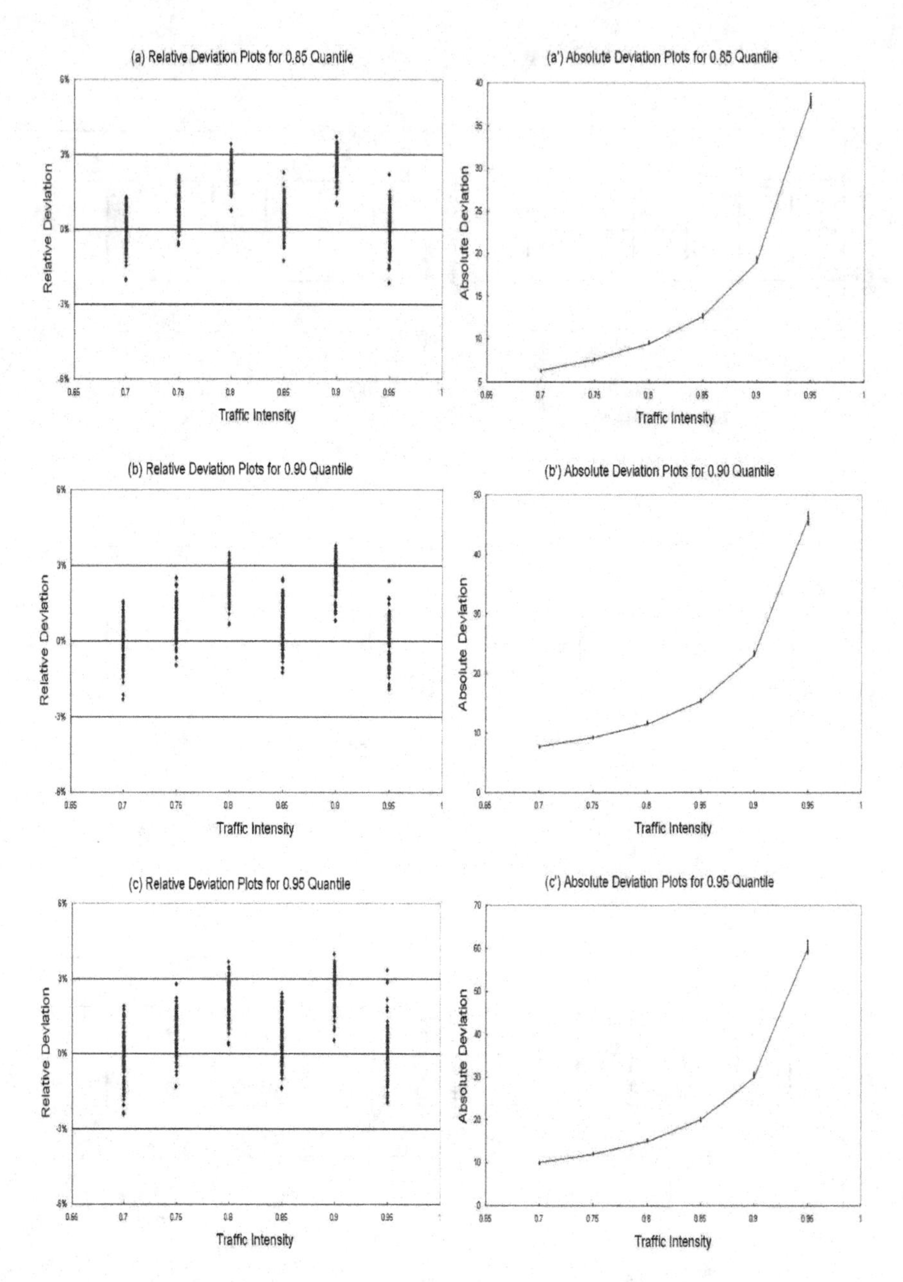

Fig. 7.3 Relative and absolute plots of the quantile estimates for M/M/1 via metamodels

Chapter 8

Density Estimation

Simulation studies have been used to investigate the characteristics of systems, for example the mean and the variance of certain system performance measures like waiting times in queue. However, without knowledge of the underlying distribution, the mean and the variance provide only limited information. On the other hand, the pdf f gives a natural description of the distribution of a stationary continuous output random variable X produced by a simulation and reveals many characteristics of the underlying distributions beyond just the mean and variance (for instance, tail probabilities and quantiles).

Density estimation from observed data is a useful tool for data exploration. For example, [Silverman (1986)] points out that "density estimates are ideal for presentation of data to provide explanation and illustration of conclusions, since they are fairly easily comprehensible to non-mathematicians." One approach to density estimation is *parametric*, assuming that the data are drawn from a known parametric family of distributions, for example the normal distribution with mean μ and variance σ^2. The density f of the underlying data is then estimated simply by estimating the values of μ and σ^2 from the data and substituting these estimates into the formula for the normal density. Another approach is *nonparametric*, where less rigid assumptions are made about the distribution of the observed data. We consider the nonparametric approach since it is more robust for the wide variety of data behavior possible in simulation output. Furthermore, the procedure is data-based, i.e., it can be embodied in a software package whose input is the simulation output data $(X_1, \ldots, X_n)$, and whose output is the density estimate. Several different approaches have received extensive treatment; see [Silverman (1986); Scott and Sain (2004)] and the references therein.

The most widely used density estimator is the *histogram*, basically a graphical estimate of the underlying probability density function that reveals all the essential distributional features of a simulation output random variable, such as skewness and multi-modality. Hence, a histogram is often used in the informal investigation of the properties of a given set of data. A steady-state distribution can be constructed with a properly selected set of quantiles. For both i.i.d. and ϕ-*mixing* sequences, sample quantiles will be asymptotically unbiased if certain conditions are satisfied; see [Sen (1972)]. With a given set of valid quantile estimates, estimating the density is more complicated than estimating the cumulative distribution because direct density estimates by central finite differences are more sensitive to the bin (band) width of the underlying histogram. This chapter investigates the performance of estimating the density of a simulation output random variable with a dynamically determined bandwidth.

8.1 Theoretical Basis

In this section, we review the definition of probability density functions and the basis of density estimation. From the definition of a probability density, if the random variable X has density f, then

$$f(x) = \lim_{b \to 0} \frac{1}{2b} P(x - b < X < x + b) = \lim_{b \to 0} \frac{F(x+b) - F(x-b)}{2b}.$$

Here $b > 0$ is a real number.

8.1.1 *Empirical Distribution Functions*

We often want to use the observed data themselves to specify a distribution, called an *empirical distribution*, from which random values are generated during the simulation. Consider an i.i.d. sequence X_i for $i = 1, 2, \cdots, n$ with distribution function F. One of the simplest and most important functions of the order statistics is the sample cumulative distribution function F_n, which can be constructed by placing a mass $1/n$ at each observation X_i. Hence, F_n may be represented as

$$F_n(x) = \frac{1}{n} \sum_{i=1}^{n} I(X_i \le x), -\infty < x < \infty.$$

Here $I(\cdot)$ is the indicator function. That is, $F_n(x)$ is the fraction of values in a sample of n values not exceeding x.

Note that

$$F_n(x) = \begin{cases} 0 & x < x_{[1]} \\ \frac{i}{n} & x_{[i]} \le x < x_{[i+1]} \\ 1 & x_{[n]} \le x. \end{cases}$$

The empirical cdf $F_n(x)$ for all real x, represents the proportion of sample values that do not exceed x. For each fixed x, the strong law of large number implies that $F_n(x) \to F(x)$ asymptotically (as the sample size n goes to ∞). The empirical distribution can then be used as a surrogate of the unknown true distribution to estimate system performance. Furthermore, the density function $f = F'$. Hence, we can use the derivative of F_n to estimate f. Since F_n is discontinuous and thus not differentiable, we consider the central finite difference

$$f_n(x) = \frac{F_n(x+b) - F_n(x-b)}{2b} = \frac{\sum_{i=1}^{n} I(x - b < X_i \le x + b)}{2nb}$$

as an estimate of the density function. The quantity $2b$ is referred to as the *bandwidth*. A graphical representation of $f_n(x)$ is called a *histogram*. Figure 8.1 shows an empirical histogram vs the density curve. The smoothness of the histogram is controlled by the parameter b. Furthermore, the starting point of each bin edge can produce different impressions of the shape, and hence a different histogram.

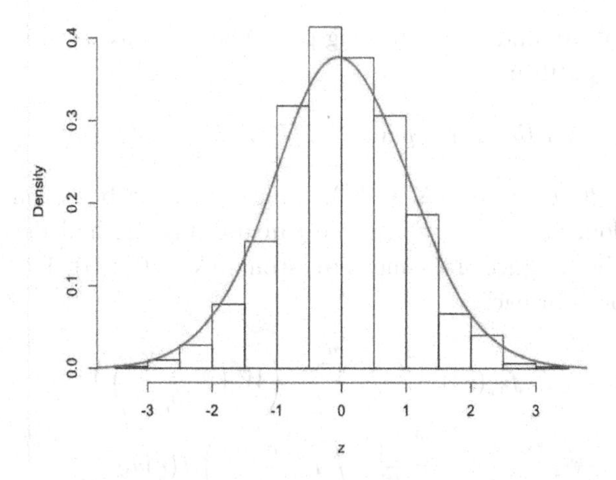

Fig. 8.1 Histogram vs density curve, bandwidth=0.5

8.1.2 *The Density Estimator*

A natural estimator by a histogram $\hat{f}_{b_n}(x)$ of the density is given by choosing a small real number b_n (which is a realization of b and depends on the sample size n) and setting

$$\begin{aligned}\hat{f}_{b_n}(x) &= \frac{1}{2nb_n}[\text{no. of } X_1,\ldots,X_n \text{ falling in } (x-b_n, x+b_n)] \\ &= \frac{1}{nb_n}\sum_{i=1}^{n} W\left(\frac{x-X_i}{b_n}\right).\end{aligned}$$

Here the weight function $W(\cdot) = I(\cdot)/2$ and $I(\cdot)$ is the indicator function for the interval $(-1, 1)$.

Let

$$I_i(b_n, x) = \begin{cases} 1 & \text{if } |x - X_i| \le b_n \\ 0 & \text{otherwise.} \end{cases}$$

The estimator $\hat{f}_{b_n}(x)$ is based on a transformation of the output sequence $\{X_i\}$ to the sequence $\{I_i(b_n, x)\}$, $i = 1, 2, \ldots, n$:

$$\hat{f}_{b_n}(x) = \frac{1}{2nb_n}\sum_{i=1}^{n} I_i(b_n, x).$$

For data that are i.i.d., the following properties of $I_i(b_n, x)$ are well known [Hogg et al. (2012)]:

$$E(I_i(b_n, x)) = p \text{ and } \mathrm{Var}(I_i(b_n, x)) = p(1-p),$$

where $p = P(-b_n < x - X < b_n)$. Since $\hat{f}_{b_n}(x)$ is based on the mean of the random variable $I_i(b_n, x)$, we can use any method developed for estimating the variance of the mean to estimate $\mathrm{Var}(\hat{f}_{b_n}(x))$. By elementary manipulations, for each x,

$$\begin{aligned}E(\hat{f}_{b_n}(x)) &= \frac{1}{nb_n}\sum_{i=1}^{n} E\left(W\left(\frac{x-X_i}{b_n}\right)\right) \\ &= \frac{1}{2b_n}\int I\left(\frac{x-y}{b_n}\right) f(y)dy \\ &= \frac{p}{2b_n}\end{aligned}$$

and

$$\text{Var}(\hat{f}_{b_n}(x)) = \frac{n}{(nb_n)^2}\text{Var}\left(W\left(\frac{x-X_i}{b_n}\right)\right)$$
$$= \frac{1}{4nb_n^2}\text{Var}\left(I\left(\frac{x-X_i}{b_n}\right)\right)$$
$$= \frac{p(1-p)}{4nb_n^2}.$$

Note that $\hat{f}_{b_n}(x)$ is a variation of a binomial distribution. It follows from the definition that $\hat{f}_{b_n}$ is not a continuous function, but has jumps at the points $X_i \pm b_n$ and has zero derivative everywhere else. This gives the estimate a somewhat ragged character.

To overcome the difficulties stemming from the ragged character of $\hat{f}_{b_n}(x)$, one can replace the weight function by a *kernel function* K, which satisfies the condition $\int_{-\infty}^{\infty} K(x)dx = 1$. For simplicity, the kernel K usually is a symmetric function satisfying $\int xK(x)dx = 0$, and $\int x^2K(x)dx = k_2 \neq 0$; an example is the normal density. The kernel density estimator with kernel K is defined by

$$\hat{f}_{K,b_n}(x) = \frac{1}{nb_n}\sum_{i=1}^{n} K\left(\frac{x-X_i}{b_n}\right).$$

The weight function W of the histogram density estimate satisfies the conditions for a Kernel function. Consequently, the histogram density estimate is just a special case of the kernel density estimate with the kernel function $K(X) = W(X) = I(X)/2$. For the remainder of this chapter, we will refer the histogram density estimate as a kernel density estimate with the indicator function as the *kernel*. For a detailed discussion of kernel density estimate, see [Silverman (1986)].

It can be shown that

$$E(\hat{f}_{K,b_n}(x)) = \frac{1}{b_n}\int K\left(\frac{x-y}{b_n}\right) f(y)dy;$$

$$\text{Var}(\hat{f}_{K,b_n}(x)) = \frac{1}{n}\left[\frac{1}{b_n^2}\int K\left(\frac{x-y}{b_n}\right)^2 f(y)dy - \left[E(\hat{f}_{K,b_n}(x))\right]^2\right]$$
$$\approx \frac{1}{nb_n} f(x)\int K(y)^2dy,$$

and

$$\text{bias}_{b_n}(x) = E(\hat{f}_{K,b_n}(x) - f(x))$$
$$= \frac{1}{2}b_n^2 f''(x)k_2 + \text{ higher-order terms in } b_n.$$

From this we can deduce that $\hat{f}_{K,b_n}(x)$ is an asymptotically unbiased estimator of the density $f(x)$, because $\text{bias}_{b_n}(x) \to 0$ when $b_n \to 0$ as $n \to 0$. See [Silverman (1986)] for details.

The approximation of bias and variance indicates one of the fundamental difficulties of density estimation. To eliminate the bias, a small value of b_n should be used, but then the variance will become large. On the other hand, a large value of b_n will reduce the variance, but will increase the bias. The mean square error (MSE) is widely used to evaluate the quality of estimates and addresses the trade-off between variance and bias. Note that MSE = Variance + Bias^2. Furthermore, to achieve MSE $\to 0$ as $n \to \infty$, the following two conditions must hold: $b_n \to 0$ and $nb_n \to \infty$. Since the shape of the true density is of most interest, a relevant criterion is the integrated mean squared error (IMSE) [Roseblatt (1971)].

There are other approaches of estimating density, e.g., [Golyandina et al. (2012)] propose using Singular Spectrum Analysis to estimate both the distribution function and the density function.

8.1.3 *The Complication of Lack of Independence*

The density estimator (obtained by the methods described above) would be asymptotically unbiased provided that the observations are independent. However, simulation output data are generally correlated and consequently the estimator maybe biased when the sample size is not sufficiently large. Hence, to estimate the density of such a stochastic processes, the procedures need to determine the sample sizes dynamically to ensure that the density estimator is unbiased. That is, we assume that the simulation output sequence satisfies the ϕ-mixing conditions. Furthermore, we assume that the underlying process is stationary; i.e., the joint distribution of the X_i's is insensitive to time shifts (in a simulation context, this would mean that the model has been adequately "warmed up").

8.2 An Implementation

This section presents a procedure to compute the point density estimator via the histogram by (four-point) Lagrange interpolation [Knuth (1998)]. Let g_i for $i = 1, 2, \ldots, G$ denote the grid points where the density estimate $\hat{f}(g_i)$ is available. For some k such that $g_{k-1} < x \leq g_k$, the density estimator at point x can be computed as follows. Let

$$\varpi_j = \prod_{j'=1, j' \neq j}^{4} \frac{x - g_{k+j'-3}}{g_{k+j-3} - g_{k+j'-3}}, \text{ for } j = 1, 2, 3, 4,$$

then $\hat{f}(x) = \sum_{j=1}^{4} \varpi_j \hat{f}(g_{k+j-3})$. In two extreme cases, $g_1 < x \leq g_2$ or $g_{G-1} < x \leq g_G$, linear interpolation will be used.

Since the procedure uses interpolation to obtain point estimates, it eliminates the ragged character of the histogram. Hence, the density estimates for different points within the same bin can have different values. Unfortunately, with interpolation the integral over the real line of the resulting function likely will not be equal to 1.

8.2.1 *Determine the Bandwidth*

[Scott and Factor (1981)] point out that "the great potential of nonparametric density estimators in data analysis is not being fully realized, primarily because of the practical difficulty associated with choosing the smoothing parameter given only data $X_1, X_2, \ldots, X_n$." There are various data-based algorithms for determining the bandwidth b_n for the kernel density estimate. [Duin (1976)] uses a modified maximum-likelihood approach, [Scott et al. (1977)] use an iterative algorithm based on an asymptotically optimal smoothing parameter. [Hearne and Wegman (1994)] use random bandwidths. [Sheather (2004)] discussed various cross-validation and plug-in methods. [Chan et al. (2010)] propose an approach for local bandwidth selection. However, these algorithms are computationally intensive.

The computationally simplest method for choosing a global bandwidth b_n is based on rules of thumb. [Silverman (1986)] suggests that the bandwidth of a general kernel estimator be $b_n = 0.9An^{-1/5}$, where A = min(standard deviation, inter-quartile range/1.34). For many purposes this will be an adequate choice of bandwidth in terms of obtaining a small IMSE. For others, it will be a good starting point for subsequent fine tuning. Let x_p be the pth sample quantile. In the procedure, we set $A = \min(\text{standard error}, (x_{0.75} - x_{0.25})/1.34)$. Let $x_{[1]}$ and $x_{[n]}$, respectively, denote the minimum and maximum of the initial n_0, $2n_0$, or $3n_0$ observations, depending on the correlation of the output sequences. Note that the sample sizes of the first three iterations are, respectively, $n_0, 2n_0$, and $3n_0$, where n_0 is the size of the QI (Quasi-Independent Sequence) buffer. The final sample size is unknown at this point and the maximum initial sample size $n = n_0, 2n_0$, or $3n_0$ is used to compute b_n. Hence, the bandwidth is likely to be larger than optimal for strongly correlated sequences because a much larger sample size will be eventually allocated.

We use the following strategy to determine the bin points. The are two categories of bins: main bins and auxiliary bins. Main bins are constructed based on the initial observations that "anchor" the bin of the simulation-generated histogram, while auxiliary bins are extensions of main bins to ensure that the bins cover future observations. The number of main bin points is $G_m = \lceil (x_{[n]} - x_{[1]})/(2b_n) \rceil$, and the number of auxiliary bin points is $G_a = 2\lceil \zeta G_m \rceil$, where $0 < \zeta < 1$. We set $\zeta = 0.1$ in the implementation. The total number of bin points is thus $G = G_m + 2G_a + 1$. Let the beginning indices of the main bin point (i.e., the origin) be $b = G_a + 1$. The procedure sets $g_{b+i} = x_{[1]} + 2ib_n$, for $i = 0, 1, \ldots, G_m + G_a - 1$, and $g_{b-i} = x_{[1]} - 2ib_n$, for $i = 1, 2, \ldots, G_a - 1$. Bin point g_1 is set to $-\infty$ and g_G is set to ∞.

It is straightforward to compute the histogram density estimator. The array $n_i, i = 2, 3, \ldots, G$ stores the number of observations between bin points g_{i-1} and g_i, so the density of $x_i = (g_{i-1} + g_i)/2$ can be estimated by $\hat{f}(x_i) = (n_i/n)/(g_i - g_{i-1})$, where $n = \sum_{i=2}^{G} n_i$ is the total number of observations. To obtain the normal kernel estimator, the procedure needs to read through the output sequence again. Because the final sample size is known, the bandwidth will be re-calculated.

8.2.2 *Determine the Sample Size*

The asymptotic validity of the density estimate is reached as the sample size or simulation run length gets large. However, in practical situations simulation experiments are restricted in time and it is not known in advance what the required simulation run length might be for the estimator to become essentially unbiased. Moreover, estimating the variance of the density estimator is needed to evaluate its precision. Therefore, a workable finite sample size must be determined dynamically for the precision required.

We use an initial sample size of $n_0 = 4000$, which is somewhat arbitrary but is tested in the empirical results below. For correlated sequences, the sample size n will be replaced with $N = nl$. Here l will be chosen sufficiently large so that systematic samples that are lag-l observations apart are essentially uncorrelated. This is possible because we assume that the underlying process satisfies the property that the autocorrelation approaches zero as the lag approaches infinity. Consequently, the final sample size N increases as the autocorrelation increases.

Since we need to process the sequence again to obtain the kernel estimator, we re-compute the bandwidth b_n with the final sample size N and the number of bin points with the new sample range. We need to allocate only the main bins because the minimum and maximum are known. Furthermore, the sample error and the quantiles $x_{0.25}$ and $x_{0.75}$ will be estimated through the histogram constructed while calculating the natural estimator. That is, the variance is conservatively estimated by

$$S_H^2 = \sum_{i=2}^{G} \max((g_{i-1} - \bar{X}(N))^2, (g_i - \bar{X}(N))^2)P_i.$$

Note that $N = nl = \sum_{i=2}^{G} n_i$, $\bar{X}(N) = \frac{1}{N}\sum_{j=1}^{N} X_j$, and $P_i = n_i/N$.

To estimate the error, the IMSE is approximated by

$$\overline{IMSE} = (2b_n/R)\sum_{r=1}^{R}\sum_{i=2}^{G-1}[\hat{f}_r(g_i) - f(g_i)]^2,$$

where R is the number of replications and $\hat{f}_r(\cdot)$ is the estimate in the rth replication. The density of g_1 and g_G is not included in the calculation because they could be $-\infty$ and ∞, respectively. Furthermore, if the true minimum $\mathbf{m}$ the true maximum $\mathbf{M}$ are known, the values $g_i < \mathbf{m}$ or $\mathbf{M} < g_i$ will not be included in the calculation.

8.2.3 *Density Confidence Interval*

Let $\hat{f}_r(x)$ denote the estimator of $f(x)$ in the rth replication. We use

$$\bar{f}(x) = \frac{1}{R}\sum_{r=1}^{R} \hat{f}_r(x)$$

as a point estimator of $f(x)$. Assuming $\bar{f}(x)$ has a limiting normal distribution, by the central limit theorem a c.i. for $f(x)$ using the i.i.d. $\hat{f}_r(x)$'s can be approximated using standard statistical procedures. That is, the ratio

$$T = \frac{\bar{f}(x) - f(x)}{S/\sqrt{R}}$$

would have an approximate t distribution with $R-1$ d.f., where

$$S^2 = \frac{1}{R-1}\sum_{r=1}^{R}\left(\hat{f}_r(x) - \bar{f}(x)\right)^2$$

is the usual unbiased estimator of the variance of $\hat{f}(x)$. This would then lead to the $100(1-\alpha)\%$ c.i., for $f(x)$,

$$\bar{f}(x) \pm t_{R-1,1-\alpha/2}\frac{S}{\sqrt{R}}, \tag{8.1}$$

where $t_{R-1,1-\alpha/2}$ is the $1-\alpha/2$ quantile for the t distribution with $R-1$ d.f. $(R \geq 2)$.

Let the half-width $w = t_{R-1,1-\alpha/2}S/\sqrt{R}$. The final step in the procedure is to determine whether the c.i. meets the user's half-width requirement, a maximum absolute half-width ϵ' or a maximum relative fraction γ of the magnitude of the final point density estimator $\bar{f}(x)$. If the relevant requirement $w \leq \epsilon'$ or $w \leq \gamma|\bar{f}(x)|$ for the precision of the confidence interval is satisfied, then the procedure terminates, returns the point density estimator $\bar{f}(x)$, and the c.i. with half-width w. If the precision requirement is not satisfied with R replications, then the procedure will increase the number of replications to

$$(w/\epsilon')^2 R \text{ or } (w/(\gamma\bar{f}(x)))^2 R. \tag{8.2}$$

This step will be executed repeatedly until the half-width is within the specified precision.

8.2.4 *The Density-Estimation Procedure*

The procedure progressively increases the simulation run length N until a pre-determined number of systematic samples (e.g. n_0) appear to be uncorrelated, as assessed by the runs test of independence. We allocate a buffer, QI (Quasi-Independent), with size $B_s = 3n_0$ to store systematic samples y_i, $1 \leq i \leq B_s$. Note that lag l' (=1,2,3) of the systematic samples is used to refer to systematic samples $y_{kl'+1}$, for $k = 0, 1, 2, \ldots, n_0 - 1$ and will be used by the runs test at various iterations.

An embedded pilot run is executed to set up the bin points. On each iteration, the algorithm operates as follows. The simulation outputs are funneled into bins. The number of observations in each bin is updated dynamically as the observation is produced during the simulation run. The systematic samples are obtained through lag-l observations and are stored in a buffer. The initial value of l is 1. Let $l' = 1, 2, 3$ denote the lag of the systematic samples stored in the buffer. If lag-l' systematic samples appear to be dependent, then the lag l is doubled every other iteration and the process is repeated until the lag-l' systematic samples appear to be independent. The initial value of l' is 0 and will be updated each iteration by the following rule: "if $l' < 3$, then $l' = l' + 1$; else $l' = 2$."

Note that l_0 is the lag used to obtain systematic samples, δ is the incremental sample size, and r is the index of iterations. Each iteration r contains two sub-iterations r_A and r_B. We limit the number of systematic samples used in the runs test to $n_0 = 4000$.

The quasi-independent-density-estimation algorithm:

(1) Initialization: Set $n_0 = 4000$, $l_0 = 1$, $\delta = n_0$, and $r = 0$.
(2) Generate δ systematic samples, which are lag-l_0 observations apart. If $r > 1$, record the number of observations in each bin.
(3) If this is the initial iteration, set $l' = 1$. If this is a r_A iteration, set $l' = 2$. If this is a r_B iteration, set $l' = 3$.
(4) Carry out the runs test to assess whether lag-l' systematic samples appear to be uncorrleated.
(5) If the lag-l' systematic samples appear to be uncorrleated, go to step 12.
(6) If $r = 0$, set $r = r + 1$ and start the 1_Ath iteration by going to step 2.
(7) If this is the 2_Ath iteration, then compute the bin points and the number of observations in each bin.
(8) If this is a r_B iteration, set $r = r + 1$ and start a r_A iteration. If this is a r_A iteration, start a r_B iteration.
(9) If this is a r_A iteration ($r > 1$), then discard the even systematic samples in the buffer, and re-index the rest of the $3n_0/2$ systematic samples in the first half of the buffer. Set $l_0 = 2^{r-1}$, $\delta = n_0/2$.
(10) If this is a r_B iteration ($r > 1$), set $\delta = n_0$.
(11) Go to step 2.
(12) If the number of replications is less than specified, go to step 1.
(13) Construct the confidence interval for $f(x)$ according to Eq. (8.1).
(14) Let ϵ' be the desired absolute half-width criterion, or let $\gamma|\bar{f}(x)|$ be the desired relative half-width criterion. If the half-width of the c.i. is greater than ϵ' or $\gamma|\bar{f}(x)|$, compute R', the required number of independent replications according to Eq. (8.2), set $R = R'$, and go to step 1; otherwise the procedure returns the c.i. estimator and terminates.

8.3 Empirical Experiments

In this section, we present some empirical results of estimating density functions with the descried method. We tested the procedure with several i.i.d. and correlated sequences. In these experiments, we used $R = 3$ independent replications to construct c.i.'s. We constructed density c.i.'s at four pre-specified points for each distribution. The confidence level $1 - \alpha$ of the density c.i. (i.e., Eq. (8.1)) is set to 0.90. Moreover, the confidence level of

Table 8.1 Coverage of 90% confidence density estimators for the $N(0, 1.81)$ distribution

avg N		4449		
stdev N		820		
x	-0.5	0	1.0	2.0
$f(x)$	0.2767	0.2965	0.2250	0.0982
Indicator Function (0.000475, 0.000228)				
coverage	89.4%	88.8%	90.3%	90.0%
avg γ	0.0168	0.0161	0.0185	0.0284
stdev γ	0.0124	0.0119	0.0140	0.0214
avg hw	0.0144	0.0145	0.0133	0.0092
stdev hw	0.0074	0.0080	0.0069	0.0051
Standard Normal (0.000278, 0.000145)				
coverage	84.1%	80.5%	87.8%	88.2%
avg γ	0.0164	0.0174	0.0149	0.0265
stdev γ	0.0119	0.0119	0.0110	0.0196
avg hw	0.0111	0.0112	0.0100	0.0073
stdev hw	0.0059	0.0059	0.0053	0.0039

the runs test for no correlation is set to (approximately) 0.90 as well. In the implementation, the runs test contains both runs up and runs down, with both tests set to $\alpha = 0.05$.

We tested the following independent sequences:

- Observations are i.i.d. from the normal distribution $N(0, 1.81)$.
- Observations are i.i.d. from the Weibull distribution $Weibull(1/2, 1)$, where $Weibull(\alpha, \beta)$ denotes the Weibull distribution with shape parameter α and scale parameter β.

Tables 8.1 and 8.2 list the experimental results using the normal and Weibull distributions, respectively. Each design point was based on 1000 replications. The *avg N* row lists the average of the sample size of each independent run. The *stdev N* row lists the standard deviation of the sample size. The x row lists the point where we want to estimate the density. The $f(x)$ row lists the true density. The values after each of the estimation methods are the $\overline{IMSE}$ and the standard error of the integrated mean squared error. The *coverage* row lists the percentage of the c.i.'s that cover the true $f(x)$. The *avg* γ row lists the average of the relative precisions of the density estimators. Here, the relative precision is defined as $\gamma = |\hat{f}(x) - f(x)|/f(x)$. The *stdev* γ row lists the standard deviation of the relative precision of the density estimators. The *avg hw* row lists the average of the c.i. half-widths. The *stdev hw* row lists the standard deviation of the c.i. half-width.

Table 8.2 Coverage of 90% confidence density estimators for the $Weibull(1/2, 1)$ distribution

avg N		4427		
stdev N		798		
x	0.5	1.0	1.5	2.0
$f(x)$	0.7358	0.2707	0.0996	0.0366
Indicator Function (0.001473, 0.000820)				
coverage	90.4%	91.0%	92.1%	89.9%
avg γ	0.0196	0.0300	0.0537	0.0915
stdev γ	0.0149	0.0223	0.0398	0.0686
avg hw	0.0466	0.0257	0.0177	0.0103
stdev hw	0.0243	0.0135	0.0088	0.0054
Standard Normal (0.000590, 0.000342)				
coverage	86.3%	88.5%	91.8%	89.6%
avg γ	0.0153	0.0256	0.0420	0.0729
stdev γ	0.0116	0.0188	0.0310	0.0552
avg hw	0.0311	0.0207	0.0133	0.0082
stdev hw	0.0168	0.0107	0.0069	0.0042

As expected, the $\overline{IMSE}$ with the standard normal as the kernel is better than with the indicator function as the kernel. Even though using the standard normal as the kernel requires more computation, the additional computational cost is minimal with today's computers. In these experiments, no relative or absolute precisions were specified, so the half-width of the c.i. is the result of the default precision. The coverages are around or slightly less than the nominal value of 90%. In general, the variance of the estimates is larger with the indicator function as the kernel. The results indicate that the variance estimates are smaller than necessary with the normal function as the kernel and result in coverage less than the nominal value. Furthermore, the estimates are likely biased low at the mode $x = 0$. With $\alpha = 0.10$, the independent sequences will fail the runs test for no correlation 10% of the time. The average sample sizes, 4449 and 4427, are close to the theoretical value, i.e., $\sum_{i=0}^{\infty} n_0 \alpha^i$, where $n_0 = 4000$.

Figures 8.2 and 8.3, respectively, show the empirical and true densities of the normal and Weibull distributions, generated from the first run of the estimate of the experiments. These figures reveal the essential characteristics of the underlying density functions. However, with the sample size just over 4000, the estimated density curves are rather ragged, especially with the indicator kernel function. As expected, the normal kernel function smoothes the ragged empirical density curve.

The density of the Weibull distribution is 0 when $x \leq 0$ and as $x > 0$ approaches 0 from the right, the density increases. The standard normal

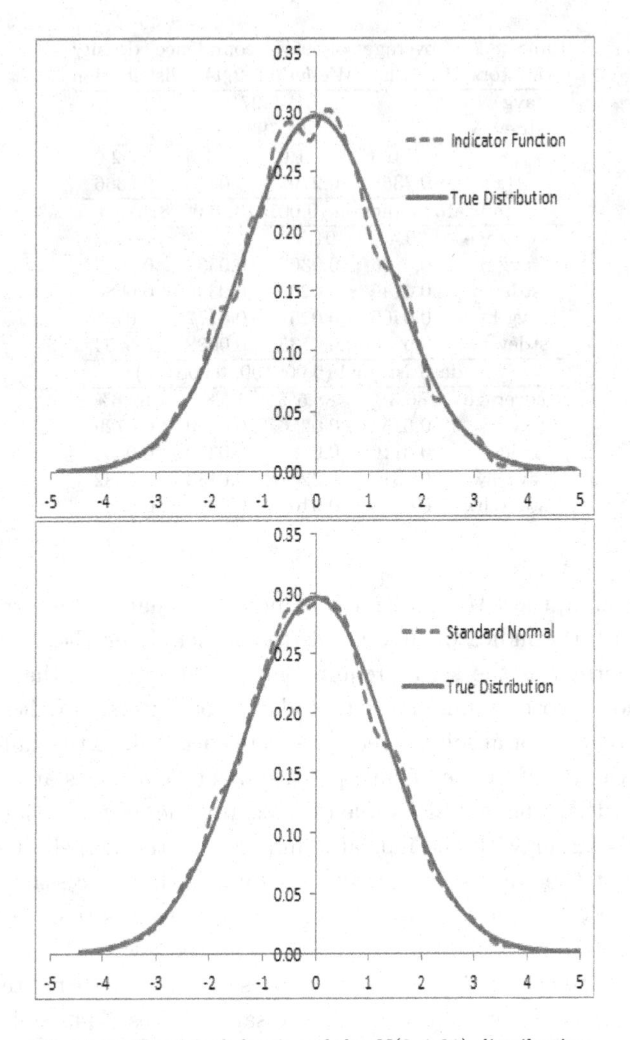

Fig. 8.2 Empirical density of the $N(0, 1.81)$ distribution

kernel density estimates over-smooths the density curve at the bounded tail and has an inflecting point. To deal with this difficulty, various adaptive methods have been proposed; see [Silverman (1986)] for more details.

We also tested the following correlated sequences:

- Observations are from the first-order moving average process $X_i = \mu + \epsilon_i + \theta\epsilon_{i-1}$ for $i = 1, 2, \ldots$, where ϵ_i is i.i.d. $N(0, 1)$ and $0 < \theta < 1$.

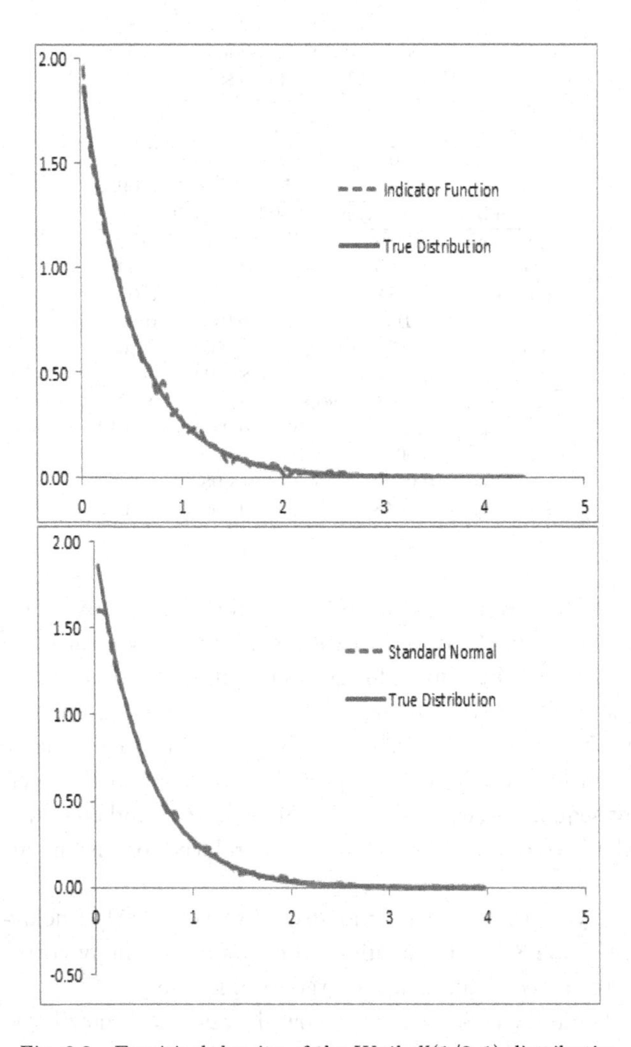

Fig. 8.3 Empirical density of the $Weibull(1/2, 1)$ distribution

This process is denoted as MA1(θ). μ is set to 0 in the experiments. It can be shown that X has an asymptotic $N(0, 1 + \theta^2)$ distribution.

- Observations are the delays in queue (exclusive of service times) from the M/M/2 queuing model, with the arrival rate $\lambda = 9$ and service rate $\nu = 5$.

For the MA1 process, we set θ to 0.90. In order to eliminate the initial

Table 8.3 Coverage of 90% confidence density estimators for the MA1(0.9) process

avg N	8457			
stdev N	864			
x	-0.5	0.0	1.0	2.0
$f(x)$	0.2767	0.2965	0.2250	0.0982
Indicator Function (0.000302, 0.000130)				
coverage	88.9%	90.2%	89.9%	90.8%
avg γ	0.0134	0.0131	0.0151	0.0234
stdev γ	0.0099	0.0097	0.0115	0.0175
avg hw	0.0118	0.0121	0.0105	0.0076
stdev hw	0.0062	0.0064	0.0054	0.0040
Normal Distribution (0.000183, 0.000090)				
coverage	83.5%	80.8%	88.7%	89.5%
avg γ	0.0135	0.0137	0.0125	0.0212
stdev γ	0.0092	0.0095	0.0094	0.0159
avg hw	0.0092	0.0093	0.0083	0.0062
stdev hw	0.0048	0.0050	0.0042	0.0032

bias, X_0 is set to a random variate drawn from the steady-state distribution $N(0, 1.81)$. Table 8.3 lists the experimental results of the MA1 process. The c.i. coverage of these four design points are around the specified 90% confidence level for both estimators. The simulation run length generally increases as the correlation coefficient θ of the MA1 process increases. The simulation run length of the MA1 process with $\theta = 0.9$ is larger than for independent sequences (with $N(0, 1.81)$ distribution) and consequently produces smaller $\overline{IMSE}$, smaller values of the relative precisions and tighter half-width.

A summary of the experimental results of the M/M/2 delay-in-queue process is in Table 8.4. The queuing processes are strongly correlated and the procedure correctly allocates a large sample size.

Figures 8.4 and 8.5, respectively, show the empirical distributions of the MA1 process with $\theta = 0.9$ and the M/M/2 delay-in-queue process with $\lambda = 9$ and $\nu = 5$, generated from the first run of the experiments. The theoretical steady-state distributions of this MA1 process and this M/M/2 queuing process are, respectively, $N(0, 1.81)$ and $1 - (81/95)e^{-x}$, where $x \geq 0$. Again, the experimental results show that these density estimates provide excellent approximations to the underlying steady-state probability density. The waiting-time density of the stationary M/M/2 delay in queue (with $\lambda = 9$ and $\nu = 5$) is $f(x) = (81/95)e^{-x}$ for $x \geq 0$ and is discontinuous at $x = 0$. Both estimators over estimate the density around the discontinuity point. With a larger sample size allocated for strongly cor-

Table 8.4 Coverage of 90% confidence density estimators for the M/M/2 process

avg N	1120128			
stdev N	224082			
x	0.5	1.0	2.0	3.0
$f(x)$	0.5171	0.3137	0.1154	0.0425
Indicator Function (0.010743, 0.009440)				
coverage	89.6%	88.3%	90.9%	89.3%
avg γ	0.0049	0.0058	0.0129	0.0250
stdev γ	0.0038	0.0044	0.0098	0.0198
avg hw	0.0079	0.0056	0.0046	0.0034
stdev hw	0.0041	0.0031	0.0024	0.0018
Standard Normal (0.000074, 0.000053)				
coverage	91.0%	90.2%	90.4%	91.9%
avg γ	0.0047	0.0057	0.0124	0.0250
stdev γ	0.0036	0.0044	0.0097	0.0198
avg hw	0.0079	0.0059	0.0047	0.0035
stdev hw	0.0043	0.0031	0.0026	0.0018

related processes, the accuracy and precision of the empirical distributions are greater. Furthermore, the bandwidth with standard normal kernel, which is calculated with the final sample size, is much smaller than with the indicator kernel.

8.4 Summary

We have evaluated a sequential procedure for estimating the density $f(x)$ of a stationary stochastic process, with or without intra-process independence. The c.i. constructed with both the indicator kernel and the normal kernel estimates obtained coverages above or around the nominal value. Because to obtain the normal kernel estimate, the procedure needs to compute the indicator kernel estimate, a prudent course is to take into account both estimates. If there are significant differences between the two estimates, further studies with large sample sizes should then be performed to investigate the cause. For example, the difference in estimates may be caused by an over-smoothed normal kernel estimate.

The indicator kernel is more suitable as a generic density estimator because it requires less computation, determines the bandwidth without the final sample size, delivers a valid c.i., and has no difficulty estimating the density around a bounded tail, though its $\overline{IMSE}$ is generally larger.

Some density estimates require more observations than others before the asymptotics necessary for density estimates to become approximately valid.

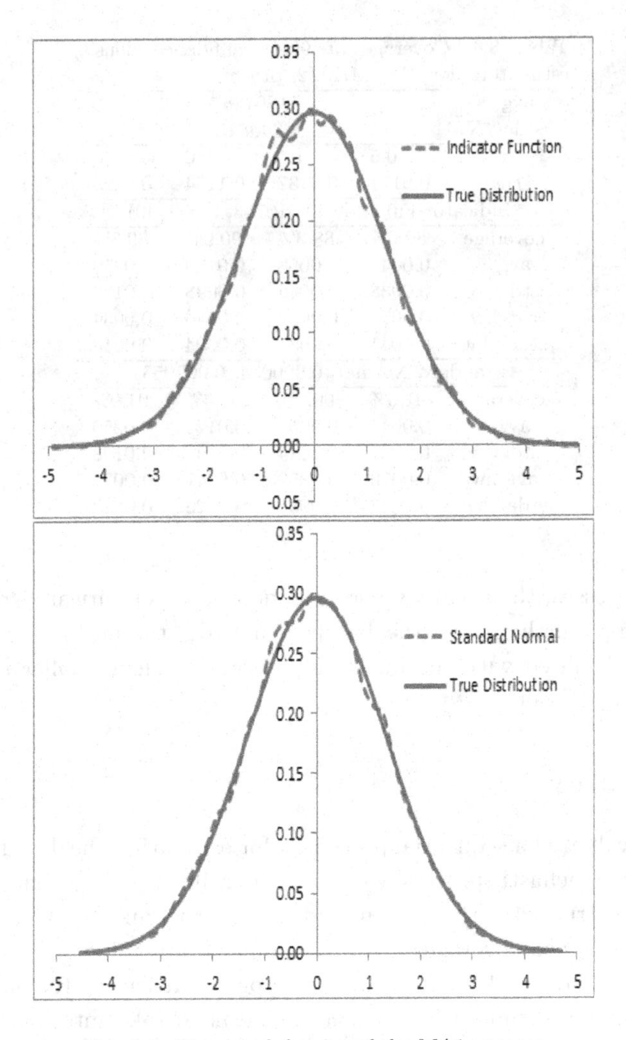

Fig. 8.4 Empirical density of the MA1 process

Our algorithm works well in determining the required simulation run length for the asymptotic approximation to become valid. The results from our empirical experiments show that the procedure is excellent in achieving the pre-specified accuracy. The procedure computes quantiles only at bin points and uses Lagrange interpolation to estimate the density at certain points. Consequently, the density at different points within the same bin can have different values. The procedure also generates an empirical distribution

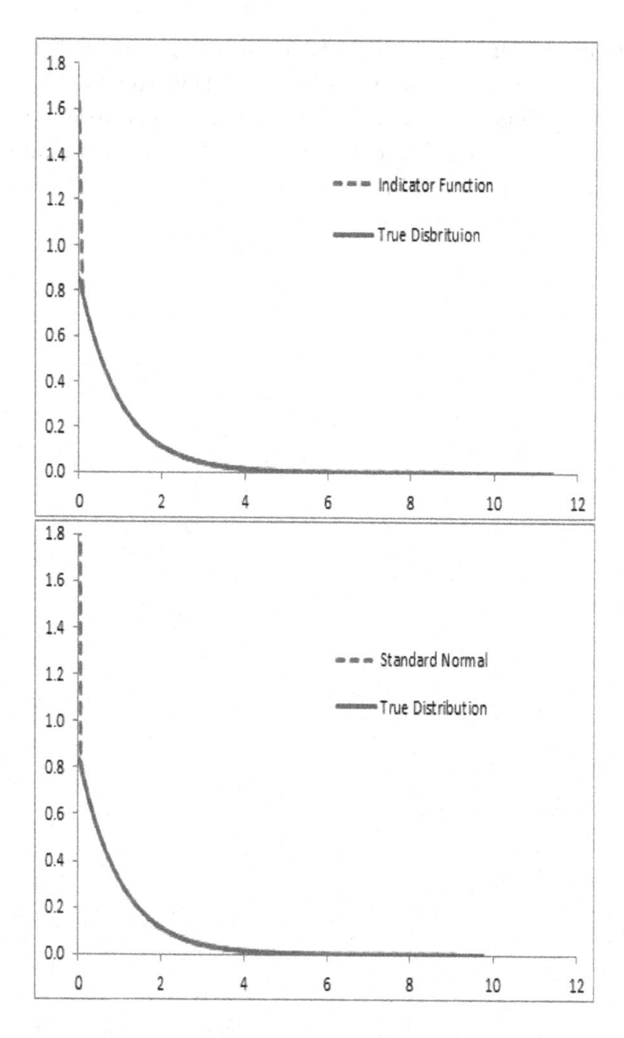

Fig. 8.5 Empirical density of the M/M/2 process

(histogram) of the output sequence, which can provide insights into the underlying stochastic process.

Our approach has the desirable properties that it is a sequential procedure and it does not require users to have *a priori* knowledge of values that the data might take on. This allows users to apply this method without having to execute a separate pilot run to determine the range of values to be expected, or guess and risk having to re-run the simulation. The main advantage of our approach is that, by using a straightforward test for lack of correlation to determine the simulation run length and obtain quantiles at bin points, we can apply classical statistical techniques directly and do not require more advanced statistical theory, thus making it easy to understand, and simple to implement.

Chapter 9

Comparing Two Alternatives

This chapter investigates the (Behrens-Fisher) problem of interval estimation and hypothesis testing concerning the difference between the means of two normally distributed populations when the variances of the two populations maybe unequal, based on two independent samples.

Let n_i denote the sample size of system i. Let X_{ij}, $i = 1, 2$ and $j = 1, 2, \ldots, n_i$ denote a sequence of mutually independent random samples from system i, and let $\mu_i = E(X_{ij})$ be the expected response of interest. We are interested in a confidence interval for $\delta = \mu_1 - \mu_2$. There are various inference procedures to construct confidence interval for δ. We propose a new approach to construct the c.i. for δ. The procedure allows the unknown variances to be unequal and allows unequal sample sizes. Furthermore, the variance reduction technique of common random numbers can be used with the procedure to reduce the confidence interval half width. We then develop a new approach to test the null hypothesis that $\mu_1 = \mu_2$.

9.1 Background

In this section, we review inference procedures of two means and the fundamentals of indifference-zone selection.

9.1.1 *Inference Procedures of Two Means*

Let $X_{ij} \sim N(\mu_i, \sigma_i^2)$. Then $X_{1j} - X_{2j} \sim N(\mu_1 - \mu_2, \sigma_1^2 + \sigma_2^2)$. Furthermore, let $\bar{X}_i = \sum_{j=1}^{n_i} X_{ij}/n_i$ for $i = 1, 2$, then $\bar{X}_1 - \bar{X}_2 \sim N(\mu_1 - \mu_2, \sigma_1^2/n_1 + \sigma_2^2/n_2)$. We review three inference procedures of constructing confidence interval of difference of $\mu_1 - \mu_2$: 1) paired-t c.i.; 2) pooled two-sample-t c.i.; 3) un-pooled two-sample-t c.i.

When constructing the paired-t c.i., the same sample sizes for each system must be equal, i.e., $n_1 = n_2 = n$. We then pair X_{1j} with X_{2j} to define $Z_j = X_{1j} - X_{2j}$, for $j = 1, 2, \ldots, n$. Then Z_j's are i.i.d. normal random variables with $E(Z_j) = \delta$. Compute sample mean

$$\bar{Z} = \frac{\sum_{j=1}^{n} Z_j}{n}$$

and sample variance

$$S^2(n) = \frac{\sum_{j=1}^{n} [Z_j - \bar{Z}]^2}{n-1}.$$

Then the $100(1-\alpha)$ percent c.i. of $\mu_1 - \mu_2$ is

$$\bar{Z} \pm t_{n-1,1-\alpha/2} S(n)/\sqrt{n}.$$

In the pooled two-sample-t approach, the sample sizes n_1 and n_2 can be different, but the variance must be equal, i.e., $\mathrm{Var}(X_{1j}) = \mathrm{Var}(X_{2j})$. Let

$$S_i^2(n_i) = \frac{\sum_{j=1}^{n_i} [X_{ij} - \bar{X}_i^2]}{n_i - 1}$$

for $i = 1, 2$. The pooled variance of X_{1j} and X_{2j} is

$$S_p^2 = \frac{(n_1 - 1)S_1^2(n_1) + (n_2 - 1)S_2^2(n_2)}{n_1 + n_2 - 2}.$$

Similarly, the pooled variance of $\bar{X}_1 - \bar{X}_2$ is $S_p^2/n_1 + S_p^2/n_2$. Note that this approach requires X_{1j}'s be independent of X_{2j}'s. Then the $100(1-\alpha)$ percent c.i. of $\mu_1 - \mu_2$ is

$$\bar{X}_1 - \bar{X}_2 \pm t_{n_1+n_2-2,1-\alpha/2} S_p \sqrt{1/n_1 + 1/n_2}.$$

For more detail, see [Devroye (1995)]. [Scheffé (1970)] points out that if $n_1 = n_2$ the pooled two-sample-t approach is fairly safe even if the variances are unequal.

In the un-pooled two-sample-t approach, the variances can be unequal, an approximate c.i. can be constructed, see [Welch (1938)]. Compute the estimated degrees of freedom

$$\hat{f} = \frac{(S_1^2(n_1)/n_1 + S_2^2(n_2)/n_2)^2}{(S_1^2(n_1)/n_1)^2/(n_1 - 1) + (S_2^2(n_2)/n_2)^2/(n_2 - 1)}.$$

Then the $100(1-\alpha)$ percent c.i. of $\mu_1 - \mu_2$ is

$$\bar{X}_1 - \bar{X}_2 \pm t_{\hat{f},1-\alpha/2} \sqrt{S_1^2(n_1)/n_1 + S_2^2(n_1)/n_2}.$$

Because $\hat{f}$ likely will not be an integer, interpolation of t tables can be used.

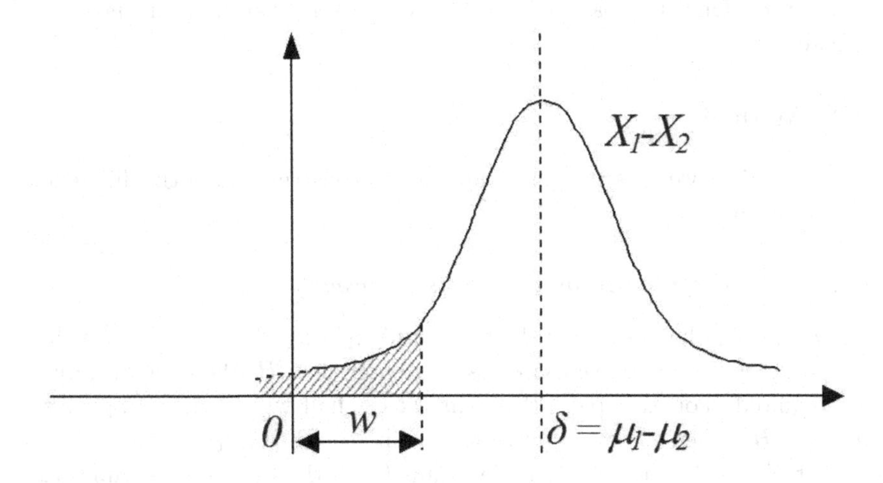

Fig. 9.1 Power of a statistical test

9.1.2 *Null Hypothesis Tests of Equivalence*

The second aspect of the Behrens-Fisher problem is testing the equality of two normal means when variances are unknown and maybe unequal. That is, the problem is to test the null hypothesis $H_0 : \mu_1 = \mu_2$ against the alternative hypothesis $H_a : \mu_1 \neq \mu_2$. While this hypothesis test can be performed by checking whether the constructed c.i. of $\mu_1 - \mu_2$ contains 0, the power of a statistical test is not addressed. Recall that in null hypothesis tests, there are Type I error with probability α that we reject the null hypothesis when it is true and Type II error with probability β that we accept the null hypothesis when it is false. The power of a statistical test is then $1 - \beta$. See Figure 9.1 for an illustration. Let the $1 - \alpha$ c.i. half width of $\delta = \mu_1 - \mu_2$ be w. The shaded area under the curve is β and the rest of the area under the curve is the power of a statistical test.

To address this issue, users need to specify two more parameters in addition to the confidence level $1-\alpha$: 1) the value of Δ, see Section 9.2.2; 2) the power of a statistical test $1-\beta$. [Dudewicz et al. (2007)] investigated three exact solutions of the Behrens-Fisher problem: 1) Dudewicz and Almed's [Dudewicz and Ahmed (1998)] procedure; 2) Chapman's [Chapman (1950)] procedure; 3) Prokof'yev and Shishkin's [Prokof'yev and Shishkin (1974)] procedure. One drawback of these procedures is that they are not user friendly.

9.2 Methodology

In this section, we present a new approach to construct c.i. of the difference of two means.

9.2.1 *A Weighted-Sample-Means Approach*

Let H denote the desired c.i. half width of $\mu_1 - \mu_2$. If the c.i. half width w_{12}, obtained by the initial samples, is greater than H, then lager samples are required. For example, if the paired-t c.i. half width with sample size n $w_{12} > H$. Then the sample size will be increased to $(w_{12}/H)^2 n$.

Let $\tilde{X}_i$ be the weighted sample means (with the first-stage sample size n_0) as defined in [Dudewicz and Dalal (1975)], see Section 4.2.1. They show that

$$T_i = \frac{\tilde{X}_i - \mu_i}{d^*/h} \text{ for } i = 1, 2$$

would have a t distribution with $n_0 - 1$ degrees of freedom. Here h is a critical constant that will be discussed later. Furthermore,

$$\vartheta = T_1 - T_2 = \frac{(\tilde{X}_1 - \tilde{X}_2) - (\mu_1 - \mu_2)}{d^*/h} = \frac{(\tilde{X}_1 - \tilde{X}_2) - (\mu_1 - \mu_2)}{\sqrt{S_1^2(n_0)/(2N_1) + S_2^2(n_0)/(2N_1)}}.$$

Here $S_i^2(n_0)$ is the unbiased estimator of the variance σ_i^2 with n_0 samples. Without loss of generality, we temporarily assume that N_1 and N_2 are real numbers. The last equality holds because

$$\frac{S_1^2(n_0)}{2N_1} = \frac{S_2^2(n_0)}{2N_2} = \frac{(d^*)^2}{2h^2} \text{ and } \sqrt{\frac{S_1^2(n_0)}{2N_1} + \frac{S_2^2(n_0)}{2N_2}} = \frac{d^*}{h}.$$

The distribution of ϑ is symmetric and its percentile (or quantile), h, can be evaluated numerically. Let

$$w_{12} = \frac{h}{\sqrt{2}} \sqrt{\frac{S_1^2(n_0)}{N_1} + \frac{S_2^2(n_0)}{N_2}}.$$

Then $w_{12} = d^*$. In practice,

$$N_i = \max(n_0 + 1, \lceil (hS_i(n_0)/d^*)^2 \rceil), \text{ for } i = 1, 2 \qquad (9.1)$$

are integers and $w_{12} \leq d^*$.

The value h is obtained such that $\Pr[T_1 - T_2 \leq h] = \int_{-\infty}^{\infty} F(t+h)f(t)dt = P^*$. Consequently, $\Pr[\vartheta \leq h] = P^*$ and $\Pr[\tilde{X}_1 - \tilde{X}_2 - d^* \leq \mu_1 - \mu_2] \geq P^*$. Note that T_i for $i = 1, 2$ are independent and the t-distribution with $n_0 - 1$ d.f. Because the distribution of $(\tilde{X}_1 - \tilde{X}_2) - (\mu_1 - \mu_2)$ has a symmetrical distribution, $\Pr[\tilde{X}_1 - \tilde{X}_2 - d^* \leq \mu_1 - \mu_2 \leq \tilde{X}_1 - \tilde{X}_2 + d^*] \geq 2P^* - 1$.

The variance of $\bar{X}_i$ is no more than the variance of $\tilde{X}_i$, consequently, $\Pr[\bar{X}_1 - \bar{X}_2 - d^* \leq \mu_1 - \mu_2 \leq \bar{X}_1 - \bar{X}_2 + d^*] \geq \Pr[\tilde{X}_1 - \tilde{X}_2 - d^* \leq \mu_1 - \mu_2 \leq \tilde{X}_1 - \tilde{X}_2 + d^*] \geq 2P^* - 1$. For some discussion of the properties of $\bar{X}$ and $\tilde{X}$, see [Chen (2011)]. Furthermore, when $\lceil (hS_i(n_0)/d^*)^2 \rceil > n_0$, $\Pr[\bar{X}_1 - \bar{X}_2 - w_{12} \leq \mu_1 - \mu_2 < \bar{X}_1 - \bar{X}_2 + w_{12}] \geq \Pr[\tilde{X}_1 - \tilde{X}_2 - w_{12} \leq \mu_1 - \mu_2 < \tilde{X}_1 - \tilde{X}_2 + w_{12}] \geq 2P^* - 1$. When $\lceil (hS_i(n_0)/d^*)^2 \rceil \leq n_0$, the weighted sample $\tilde{X}_i$ purposely loses some information so that the required c.i. half width is d^*.

The interval $\bar{X}_1 - \bar{X}_2 \pm w_{12}$ is a valid c.i. for $\mu_1 - \mu_2$ with tight half-width $w_{12} \leq d^*$. However, when the goal is to test the null hypothesis $\mu_1 = \mu_2$, the weighted sample means should be used because they have the same distribution.

9.2.2 *Fix the Value of $\beta = \alpha/2$ of Null Hypothesis Tests*

Our solution of the Behrens-Fisher problem is similar to that of [Chapman (1950)]. To make the approach user friendly, we fix the value of $\beta = \alpha/2$. The test at confidence level $1 - \alpha$ of $H_0 : \mu_1 = \mu_2$ against the alternative $H_a : \mu_1 \neq \mu_2$ is based on the test statistic

$$\vartheta = T_1 - T_2.$$

The acceptance region for this test is $|\vartheta| \leq h_{1-\alpha/2}$, where $h_{1-\alpha}$ is the $1-\alpha$ quantile of the distribution of ϑ. That is,

$$|\tilde{X}_1 - \tilde{X}_2| \leq \frac{h_{1-\alpha/2}}{\sqrt{2}} \sqrt{\frac{S_1^2(n_0)}{N_1} + \frac{S_2^2(n_0)}{N_2}}.$$

Let

$$Y = \frac{\tilde{X}_1 - \tilde{X}_2 - \Delta}{\sqrt{\frac{S_1^2(n_0)}{2N_1} + \frac{S_2^2(n_0)}{2N_2}}}.$$

When the the difference of the two means $\Delta = |\delta|$, the probability of committing a Type II error, i.e., concluding that the null hypothesis is true when in fact it is false, is

$$\beta = \Pr[-h_{1-\alpha/2} - \frac{\Delta}{\sqrt{\frac{S_1^2(n_0)}{2N_1} + \frac{S_2^2(n_0)}{2N_2}}} \le Y \le h_{1-\alpha/2} - \frac{\Delta}{\sqrt{\frac{S_1^2(n_0)}{2N_1} + \frac{S_2^2(n_0)}{2N_2}}}]$$

$$= G(h_{1-\alpha/2} - \frac{\Delta}{\sqrt{\frac{S_1^2(n_0)}{2N_1} + \frac{S_2^2(n_0)}{2N_2}}}) - G(-h_{1-\alpha/2} - \frac{\Delta}{\sqrt{\frac{S_1^2(n_0)}{2N_1} + \frac{S_2^2(n_0)}{2N_2}}})$$

$$\le G(h_{1-\alpha/2} - \frac{\Delta}{\sqrt{\frac{S_1^2(n_0)}{2N_1} + \frac{S_2^2(n_0)}{2N_2}}}).$$

Here G is the distribution function of ϑ. Furthermore, the value of

$$G(-h_{1-\alpha/2} - \frac{\Delta}{\sqrt{\frac{S_1^2(n_0)}{2N_1} + \frac{S_2^2(n_0)}{2N_2}}})$$

will be very small, i.e., much smaller than β.

For fixed Δ and α, β can be evaluated as a function of sample sizes N_1 and N_2. Suppose we want to limit the probability of β, the sample sizes N_1 and N_2 should be large enough such that

$$h_{1-\alpha/2} - \frac{\Delta}{\sqrt{\frac{S_1^2(n_0)}{2N_1} + \frac{S_2^2(n_0)}{2N_2}}} = h_\beta.$$

Note that the true probability of committing Type II error will be less than β. The equation can be achieved when

$$\frac{N_1}{S_1^2(n_0)} = \frac{N_2}{S_2^2(n_0)} = (\frac{h_{1-\alpha/2} - h_\beta}{\Delta})^2.$$

That is, the sample size is $N_i = \lceil (h_{1-\alpha/2} - h_\beta)^2 S_i^2(n_0)/\Delta^2 \rceil$ for $i = 1, 2$.

Let $\beta = \alpha/2 = 1 - P^*$. Recall that $h_{1-\alpha/2} = -h_{\alpha/2}$. The sample size for system i is then computed by

$$N_i = \max(n_0 + 1, \lceil (2h_{1-\alpha/2} S_i(n_0)/\Delta)^2 \rceil). \tag{9.2}$$

This is the same as Eq. (9.1) with $d^* = \Delta/2$. Similarly, the weight W_{i1} of Eq. (4.3) needs to computed with $d^* = \Delta/2$. When $N_i > n_0$, $w_{12} = (h_{1-\alpha/2}/\sqrt{2})\sqrt{S_1^2(n_0)/N_1 + S_2^2(n_0))/N_2} \le \Delta/2$. If $\mu_1 = \mu_2$, then $\Pr[|\tilde{X}_1 - \tilde{X}_2| < w_{12}] \ge 1 - \alpha$. If $|\mu_1 - \mu_2| \ge \Delta$, then $\Pr[|\tilde{X}_1 - \tilde{X}_2| < w_{12}] \le \beta$.

With this approach, the critical constant $h_{1-\alpha/2}$ does not depend on Δ and a simpler critical constant table can be used. We believe this can increase the usability of the procedure. Note that the value of $|\delta|$ is unknown. Nevertheless, if the specified $\Delta \le |\delta|$, then the probability of committing Type II error will be than less β.

Table 9.1 lists the critical constant h, which depends on α and n_0.

Table 9.1 Values of critical constant h

n_0/α	0.010	0.025	0.050	0.100	0.250
2	9.9998	9.9996	9.9992	6.1568	2.0000
3	9.9991	6.5399	4.5661	3.0405	1.3663
4	6.4562	4.6400	3.5122	2.5045	1.2039
5	5.1889	3.9624	3.1152	2.2919	1.1358
6	4.6282	3.6389	2.9144	2.1780	1.0965
7	4.3182	3.4518	2.7942	2.1074	1.0711
8	4.1232	3.3304	2.7145	2.0596	1.0533
9	3.9897	3.2454	2.6578	2.0250	1.0401
10	3.8925	3.1827	2.6155	1.9989	1.0301
12	3.7613	3.0966	2.5566	1.9620	1.0156
14	3.6769	3.0402	2.5176	1.9373	1.0058
16	3.6181	3.0005	2.4898	1.9196	0.9986
18	3.5747	2.9710	2.4691	1.9063	0.9932
20	3.5415	2.9483	2.4531	1.8959	0.9890
22	3.5152	2.9302	2.4403	1.8876	0.9855
24	3.4940	2.9155	2.4298	1.8808	0.9827
26	3.4764	2.9033	2.4211	1.8751	0.9804
28	3.4615	2.8930	2.4138	1.8703	0.9784
30	3.4489	2.8842	2.4075	1.8662	0.9766
32	3.4380	2.8766	2.4020	1.8626	0.9751
34	3.4285	2.8700	2.3973	1.8595	0.9738
36	3.4202	2.8641	2.3931	1.8567	0.9727
38	3.4128	2.8589	2.3893	1.8543	0.9717
40	3.4062	2.8543	2.3860	1.8521	0.9707
42	3.4003	2.8502	2.3830	1.8501	0.9699
44	3.3950	2.8464	2.3803	1.8483	0.9691
46	3.3901	2.8430	2.3778	1.8467	0.9684
48	3.3857	2.8399	2.3756	1.8452	0.9678
50	3.3817	2.8370	2.3735	1.8438	0.9672

9.3 Empirical Experiments

In this section, we present some empirical results.

9.3.1 *Experiment 1: Difference of Means*

In this experiment, the unknown means for all systems are $\mu_i = 0$ and the unknown variances of system i σ_i^2 for $i = 1, 2$. Table 9.2 lists the configuration of variances. The required parameter d^* is set to 0.2 or 0.6, which implies that the targeted one-tailed confidence interval half-width is 0.2 or 0.6, respectively. The sample size for each system is computed according to Eq. (9.1). The initial sample size $n_0 = 20$. We experimented with the following nominal confidence levels 0.95, 0.975, and 0.99, hence, the critical constant $h = 2.4531, 2.9483$, and 3.5415. The resulting two-

Table 9.2 Variance configuration

Model	σ_1^2, σ_2^2
Equal	1, 1
Increasing	1, $1 + d^*$
Decreasing	1, $\frac{1}{1+d^*}$

Table 9.3 Percentage of the coverage with equal means and $d^* = 0.2$

Configuration	$1-\alpha$	half width	0.90	0.95	0.98
	Weighted	d^*	0.9033	0.9472	0.9802
	Overall	d^*	0.9053	0.9474	0.9801
Equal	Weighted	w_{12}	0.9028	0.9468	0.9801
	Overall	w_{12}	0.9050	0.9471	0.9799
	$\bar{T}$		302	434	626
	Weighted	d^*	0.9027	0.9533	0.9789
	Overall	d^*	0.9028	0.9534	0.9790
Increasing	Weighted	w_{12}	0.9021	0.9527	0.9787
	Overall	w_{12}	0.9024	0.9527	0.9790
	$\bar{T}$		331	479	690
	Weighted	d^*	0.8984	0.9504	0.9795
	Overall	d^*	0.8977	0.9507	0.9790
Decreasing	Weighted	w_{12}	0.8980	0.9500	0.9793
	Overall	w_{12}	0.8972	0.9507	0.9789
	$\bar{T}$		276	399	577

tailed c.i. will have 0.90, 0.95, and 0.98 confidence. We list the percentage that the c.i. (constructed four different ways) contain the true differences $\mu_1 - \mu_2$ (i.e., $0 \in \tilde{X}_1 - \tilde{X}_2 \pm d^*$, $0 \in \bar{X}_1 - \bar{X}_2 \pm d^*$, $0 \in \tilde{X}_1 - \tilde{X}_2 \pm w_{12}$, $0 \in \bar{X}_1 - \bar{X}_2 \pm w_{12}$).

Tables 9.3 and 9.4 list the results with $d^* = 0.2$ and 0.6, respectively. In addition to the observed coverages, we also list the average sample size of each simulation run $\overline{T}$, i.e., $\overline{T} = \sum_{r=1}^{10000} \sum_{i=0}^{k} N_{r,i}/10000$, $N_{r,i}$ is the total number of replications or batches for system i in the rth independent run. The procedure correctly increases and decreases the allocated sample sizes as the variance increases and decreases, respectively. The observed coverages of the c.i. constructed by $0 \in \tilde{X}_1 - \tilde{X}_2 \pm d^*$ and $0 \in \bar{X}_1 - \bar{X}_2 \pm w_{12}$ are close to the nominal values. Because the overall sample mean $\bar{X}_i$ has smaller variance than the weighted sample mean $\tilde{X}_i$, the coverages of the confidence intervals built by $\bar{X}_i$ is generally greater than those built by $\tilde{X}_i$, with the same half width.

In the setting that $d^* = 0.6$, the initial sample size (i.e., 42= (20+1)2) is large enough to achieve greater precision than the specified precision in two

Table 9.4 Percentage of the coverage with equal means and $d^* = 0.6$

Configuration	$1-\alpha$	half width	0.90	0.95	0.98
	Weighted	d^*	0.8973	0.9484	0.9776
	Overall	d^*	0.9492	0.9619	0.9808
Equal	Weighted	w_{12}	0.8413	0.9317	0.9754
	Overall	w_{12}	0.9046	0.9503	0.9792
	$\bar{T}$		42	52	70
	Weighted	d^*	0.9047	0.9461	0.9791
	Overall	d^*	0.9346	0.9574	0.9812
Increasing	Weighted	w_{12}	0.8704	0.9377	0.9768
	Overall	w_{12}	0.9090	0.9497	0.9801
	$\bar{T}$		49	64	91
	Weighted	d^*	0.8995	0.9532	0.9788
	Overall	d^*	0.9685	0.9756	0.9850
Decreasing	Weighted	w_{12}	0.7995	0.9203	0.9706
	Overall	w_{12}	0.9053	0.9555	0.9803
	$\bar{T}$		42	47	59

cases. The weighted sample means purposely lose information so that the observed coverages of the c.i. constructed by $\tilde{X}_1 - \tilde{X}_2 \pm d^*$ are close to the nominal values. On the other hand, in those cases the observed coverages of the c.i. constructed by $\tilde{X}_1 - \tilde{X}_2 \pm w_{12}$ are less than the nominal values because $N_i = n_0 + 1$ for $i = 1, 2$ are greater than the required sample sizes (say a_i for $i = 1, 2$) in two cases. Hence,

$$w_{12} = \frac{h}{\sqrt{2}}\sqrt{\frac{S_1^2(n_0)}{N_1} + \frac{S_2^2(n_0)}{N_2}} \le \frac{h}{\sqrt{2}}\sqrt{\frac{S_1^2(n_0)}{a_1} + \frac{S_2^2(n_0)}{a_2}} \approx d^*.$$

That is, in the cases that $a_i \le n_0$, the variance of $\tilde{X}_i$ is greater than $\bar{X}_i$ while the c.i. half width w_{12} is computed based on the variance of $\bar{X}_i$ and results in low coverages. On the other hand, the coverages of c.i. constructed by $\bar{X}_1 - \bar{X}_2 \pm d^*$ are greater than the nominal values. i.e., the half width is wider than necessary.

9.3.2 *Experiment 2: Null Hypothesis of Equal Means*

In this experiment, the unknown means have two settings: 1) $\mu_1 = \mu_2 = 0$; 2) $\mu_1 = 0$ and $\mu_2 = d^*$. The configuration of variances is as before, i.e., Table 9.2. The required parameter Δ is set to 0.2, which implies that the targeted one-tailed confidence interval half-width is $d^* = \Delta/2 = 0.1$. The sample size for each system is computed according to Eq. (9.2). The nominal confidence level is $1 - \alpha = 0.90, 0.95$, and 0.98. We list the percentage of P(CD) (i.e., the probability of correct decision). That is, in

Table 9.5 Percentage $|\tilde{X}_1 - \tilde{X}_2| < w_{12}$ with equal means

Configuration	$1-\alpha$	0.90	0.95	0.98
Equal	P(CD)	0.9011	0.9516	0.9808
	$\bar{T}$	1203	1736	2519
Increasing	P(CD)	0.9000	0.9526	0.9813
	$\bar{T}$	1323	1916	2770
Decreasing	P(CD)	0.9008	0.9488	0.9795
	$\bar{T}$	1105	1587	2303

Table 9.6 Percentage $|\tilde{X}_1 - \tilde{X}_2| \geq w_{12}$ with unequal means

Configuration	$1-\beta$	0.95	0.975	0.99
Equal	P(CD)	0.9519	0.9733	0.9895
	$\bar{T}$	1203	1736	2519
Increasing	P(CD)	0.9507	0.9770	0.9915
	$\bar{T}$	1323	1916	2770
Decreasing	P(CD)	0.9500	0.9733	0.9900
	$\bar{T}$	1105	1587	2303

the first setting, we list the percentage that $|\tilde{X}_1 - \tilde{X}_2| < w_{12}$, i.e., we don't reject the null hypothesis that $\mu_1 = \mu_2$. In the second setting, we list the percentage that $|\tilde{X}_1 - \tilde{X}_2| \geq w_{12}$, i.e., we reject the null hypothesis that $\mu_1 = \mu_2$.

Tables 9.5 and 9.6 list the results of settings 1 and 2, respectively. The observed P(CD)'s are close to the nominal values. The procedure correctly allocates the sample sizes according to the variances and is effective in terms of Type I and Type II errors.

9.4 Summary

We have presented a new approach to construct c.i. of the difference of two means as well as performing the null hypothesis test of equivalence that controls the probability of Type II error. The approach allows the unknown variances to be unequal and allow unequal sample sizes. Because the required critical constant does not involve the parameter Δ, the table of critical constants is simplified, which can increase the useability of the procedure. Furthermore, common random numbers can be used with the procedure to reduce the half width and the range of the c.i.

Chapter 10

Ranking and Selection

Among many other aspects, reliability analysis studies the expected life and the failure rate of a component or a system of components linked together in some structure. We propose applying ranking-and-selection procedures to this analysis. That is, we are more interested in whether a given component is better than the others rather than the accuracy of the performance measures. The underling philosophy is to rank estimators through ordinal comparison while the precision of the estimates are still poor [Ho et al. (1992)], hence, increase the efficiency of reliability analysis. A detailed analysis to obtain more accurate performance measures can then be carried out on the few chosen systems.

When evaluating k alternative system designs, one or more systems are selected as the best; and the probability that the selected systems really are the best is controlled. Let μ_i denote the expected response of system i and let μ_{i_l} denote the lth smallest of the μ_i's such that $\mu_{i_1} \leq \mu_{i_2} \leq \ldots \leq \mu_{i_k}$. The goal is to select a subset of size m containing the v best of k systems. We derive the probability lower bound of correctly selecting a subset based on the distribution of order statistics in a clear and concise manner. If $m = v = 1$, then the problem is to choose the best system. When $m > v = 1$, we are interested in choosing a subset of size m containing the best. If $m = v > 1$, we are interested in choosing the m best systems.

Many selection procedures are derived based on the least favorable configuration (LFC), i.e., assuming $\mu_{i_1} = \mu_{i_2} = \cdots = \mu_{i_v}$ and $\mu_{i_v} + d^* = \mu_{i_{v+1}} = \cdots = \mu_{i_k}$. This is because the minimal P(CS), or the probability of Correct Selection, occurs under the LFC. If the difference $\mu_{i_{v+1}} - \mu_{i_v} < d^*$, then these systems are considered to be in the indifference zone for correct selection. On the other hand, if the difference $\mu_{i_{v+1}} - \mu_{i_v} \geq d^*$, then these systems are considered to be in the preference zone for correct selection. Indifference-zone selection procedures attempt to avoid making a large number of replications or batches to resolve differences less than d^*. The goal is to make a correct selection with a probability of at least P^* provided that $\mu_{i_{v+1}} - \mu_{i_v} \geq d^*$.

We present a framework for indifference-zone selection that is applicable for the normal and lognormal populations. This framework is derived based on the distribution of order statistics.

10.1 Introduction

First, some notations:

X_{ij}: the observations from the jth replication or batch of the ith system;
n_0: the number of initial replications for all systems of multi-stage procedures;
n_i: the number of replications for system i, where $i \geq 1$;
μ_i: the expected performance measure for system i, i.e., $\mu_i = E(X_{ij})$;
$\bar{X}_i(n_0)$: the sample mean performance measure for system i, i.e., $\bar{X}_i(n_0) = \sum_{j=1}^{n_0} X_{ij}/n_0$;
$\bar{X}_i$: shorthand for $\bar{X}_i(n_i)$, i.e., the sample means with all samples of system i;
σ_i^2: the variance of the observed performance measure of system i from one replication or batch, i.e., $\sigma_i^2 = \text{Var}(X_{ij})$;
$S_i^2(n_i)$: the sample variance of system i with n_i replications, i.e., $S_i^2(n_i) = \sum_{j=1}^{n_i}(X_{ij} - \bar{X}_i)^2/(n_i - 1)$.

We derive the subset selection procedure and extend the procedure to select the best system when the parameter of interest is variance or the underlying populations are lognormally distributed.

10.1.1 *Generalized Subset Selection*

[Mahamunulu (1967)] considers a generalized version of the selection problem. The objective is to select a subset of size m containing at least c of the v best of k systems, where $\max(1, m + v + 1 - k) \leq c \leq \min(m, v)$ and $\max(m, v) \leq k - 1$. Furthermore, the minimum required P(CS) satisfies the following:

$$P^* \geq P(c, v, m, k) = \binom{k}{m}^{-1} \sum_{i=c}^{\min(m,v)} \binom{v}{i} \binom{k-v}{m-i}.$$

If $P^* < P(c, v, m, k)$, then the precision requirement can be achieved by choosing the subset at random. In the case that $c = v = m = 1$, the goal is to select the best of k systems.

Selection procedures generally simulate n_i samples for system i and rank the sample means such that $\bar{X}_{b_1} \leq \bar{X}_{b_2} \leq \cdots \leq \bar{X}_{b_k}$ and select systems b_l for $l = 1, 2, \ldots, m$ as the best m systems. P(CS) of this selection problem is now assessed. Let $\bar{X}_{[c]}$ be the cth smallest sample mean from $\bar{X}_{i_l}$ for $l = 1, 2, \ldots, v$ and let $\mu_{[c]}$ be its unknown true mean. Let $\bar{X}_{[u]}$ ($u = m-c+1$) be the smallest sample mean from $\bar{X}_{i_l}$ for $l = v+1, v+2, \ldots, k$ and let $\mu_{[u]}$ be its unknown true mean. To simplify the notation, we will use $\bar{X}_c$, $\bar{X}_u$, μ_c, and μ_u instead of $\bar{X}_{[c]}$, $\bar{X}_{[u]}$, $\mu_{[c]}$, and $\mu_{[u]}$ in the remainder of this chapter. Then correct selection occurs if and only if $\bar{X}_c < \bar{X}_u$. An important point is that systems c and u are unknown.

Let w_{cu} be the one-tailed P^* c.i. half-width of $\bar{X}_c - \bar{X}_u$. Then

$$\begin{aligned}
\mathrm{P(CS)} &= \Pr[\bar{X}_c < \bar{X}_u] \\
&= \Pr[\mu_u - \mu_c - d^* < \bar{X}_u - \bar{X}_c] \\
&\geq \Pr[\mu_u - \mu_c - w_{cu} < \bar{X}_u - \bar{X}_c] \\
&= \Pr[\mu_u - \mu_c < \bar{X}_u - \bar{X}_c + w_{cu}] \\
&= P^*.
\end{aligned}$$

The second equality holds because $\mu_u - \mu_c = d^*$ under the LFC. The inequality holds because the procedure will allocate large enough sample sizes such that $w_{cu} \leq d^*$. The last equality holds from the definition of the c.i.

10.1.2 *Order Statistics of Continuous Distributions*

Let $f(\cdot|\theta)$ and $F(\cdot|\theta)$, respectively, denote pdf and cdf of the random variable Y given a parameter θ, e.g., the mean of a normal distribution. Let y be a realization of Y. From Chapter 4, the distribution of the uth order statistics of k observations of Y is

$$g_{u:k}(y|\theta) = \beta(F(y|\theta); u, k-u+1) f(y|\theta).$$

Let

$$I(p; a, b) = \int_0^p \beta(x; a, b) dx, \text{ where } a, b > 0.$$

Assuming $E[f(\cdot|\theta_c)] = \mu_{i_c}$ and $E[f(\cdot|\theta_u)] = \mu_{i_u}$. Let $T_c \sim g_{c:v}(\cdot|\theta_c)$ and $T_u \sim g_{m-c+1:k-v}(\cdot|\theta_u)$. Then,

$$\begin{aligned}
&\Pr[T_c - T_u \leq h] \\
&= \int_{-\infty}^{\infty}\int_{-\infty}^{y+h} g_{c:v}(x|\theta_c)g_{m-c+1:k-v}(y|\theta_u)dxdy \\
&= \int_{-\infty}^{\infty} G_{c:v}(y+h|\theta_c)dG_{m-c+1:k-v}(y|\theta_u) \\
&= \int_{-\infty}^{\infty} I(F(y+h|\theta_c), c, v-c+1)dI(F(y|\theta_u), m-c+1, k-v).
\end{aligned}$$

The right-hand side is then equated to P^* to solve for h. Values of the critical constant h can be found in Table A.1. When $h = 0$ the equation is the same as Eq. (4.14) in [Mahamunulu (1967)].

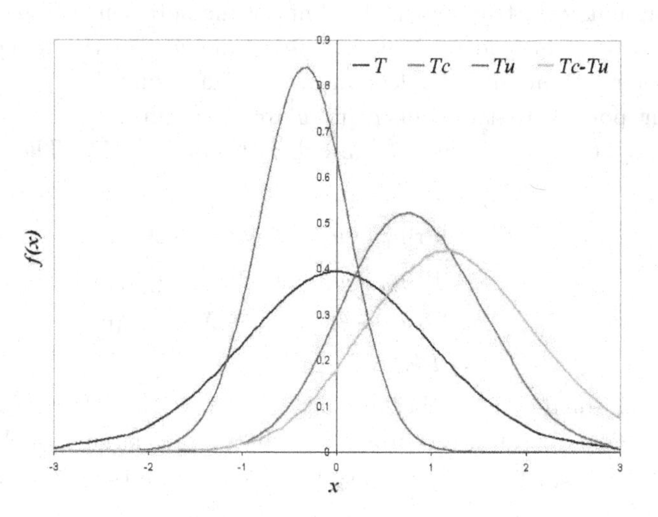

Fig. 10.1 Empirical probability densities of T

For example, if we are interested in the probability of correctly selecting a subset of size 5 containing 3 of the first 3 best from 10 alternatives, then $T_c \sim g_{3:3}(t_c)$ and $T_u \sim g_{3:7}(t_u)$. Furthermore, if the initial sample size is $n_0 = 20$, then f and F are, respectively, the pdf and cdf of the t-distribution with 19 d.f. Figure 10.1 displays the pdf of the random variables T having a t-distribution with 19 d.f., T_c, T_u, and $T_c - T_u$.

Note that the value of h is determined such that $\Pr[T_c - T_u \leq h] = P^*$. Let $\vartheta = T_c - T_u$, then $\Pr[\vartheta < h] = P^*$. That is, under the LFC the value of h is the P^* quantile of the distribution of ϑ. This property can be used to estimate the value of h for the problem at hand.

The absolute indifference zone (i.e., $0 < \mu_{i_{v+1}} - \mu_{i_v} < d_a^* = d^*$) is generally applied when the performance measure of interest is a location parameter. When the performance measure of interest is a scale parameter, the relative indifference zone (i.e., $1 \leq \mu_{i_{v+1}}/\mu_{i_v} < d_r^*$) is generally applied. In the case of the relative indifference zone

$$\begin{aligned}&\Pr[T_c/T_u \leq h]\\&= \int_{-\infty}^{\infty} I(F(yh|\theta_c), c, v-c+1) dI(F(y|\theta_u), m-c+1, k-v).\end{aligned}$$

The right-hand side is then equated to P^* to solve for h. Note that the value of h is determined such that $\Pr[\varphi < h] = P^*$, where $\varphi = T_c/T_u$.

In cases where the parameter θ is either a location parameter or a scale parameter, the procedure needs only the value of d_a^* or d_r^* to determine the required sample size. On the other hand, the procedure needs the values of θ_c and θ_u to determine the required sample sizes, when θ is neither a location parameter nor a scale parameter and $\theta_c \neq \theta_u$.

10.1.3 *A Review of Confidence Interval Half Width*

Let X_i be a normally distributed random variable with mean μ_i and variance σ_i^2. Let z_p denote the p quantile of the standard normal distribution. It is known that

$$\Pr\left[\bar{X}_i - w_i \leq \mu_i\right] = p.$$

Here $w_i = z_p \sigma_i/\sqrt{n_i}$ is the one-tailed p c.i. half width of μ_i. Furthermore, by the symmetry of the normal distribution

$$\Pr\left[\mu_i \leq \bar{X}_i + w_i\right] = p.$$

Under the LFC, $\mu_{i_1} = \mu_{i_2} = \cdots = \mu_{i_v}$ and $\mu_{i_{v+1}} = \mu_{i_{v+2}} = \cdots = \mu_{i_k}$. We consider the case that $\sigma_i^2 = \sigma^2$ and $n_i = n_0$ for $i = 1, 2, \ldots, k$. Let

$$Z_i = \frac{\bar{X}_i - \mu_i}{\sigma/\sqrt{n_0}} \text{ for } i = 1, 2, \ldots, k.$$

Let Z_c be the cth order statistics of the Z_i for $i = 1, 2, \ldots, v$. Note that Z_c is no longer normally distributed. Let h_x be a critical constant such that

$$\Pr\left[Z_c \leq h_x\right] = p.$$

Then $\Pr\left[\bar{X}_c - w_c \leq \mu_c\right] = p$. Note that h_x is the p quantile of the distribution of $(\bar{X}_c - \mu_c)/(\sigma/\sqrt{n_0})$ and $w_c = h_x \sigma/\sqrt{n_0}$ is the one-tailed p c.i. half width of μ_c. Note that the value of σ is a scale parameter of the distribution of $\bar{X}_c$. Moreover, based on the primitive study, $\bar{X}_c$ has a symmetric distribution so that $\Pr\left[\mu_c \leq \bar{X}_c + w_c\right] = p$.

Let Z_u be the uth order statistics of the Z_i for $i = v+1, v+2, \ldots, k$. It is known if $\Pr\left[\bar{X}_c - w_c \leq \mu_c\right] = p$ and $\Pr\left[\mu_u \leq \bar{X}_u + w_u\right] = p$, then $\Pr\left[\bar{X}_c - \bar{X}_u \leq \mu_c - \mu_u + w_c + w_u\right] = 2p-1$. Hence, if $w_c + w_u \leq \mu_u - \mu_c = d^*$, then $\Pr\left[\bar{X}_c \leq \bar{X}_u\right] \geq 2p - 1$. Let $p = (1 + P^*)/2$. To achieve the specified probability guarantee, the sample sizes should be large enough such that w_c and w_u are less than $d^*/2$. That is, $n_c = \lceil (2h_x\sigma_c/d^*)^2 \rceil$ and $n_u = \lceil (2h_y\sigma_u/d^*)^2 \rceil$. Here h_y is a critical constant similar to h_x.

We would like to point out that, when the underlying distributions are non-normal, there may not be a scale parameter and the c.i. half width may be dependent upon other distribution parameters, such as location or shape parameters.

10.1.4 *Confidence Interval Half Width of Interest*

In the previous section, we show how to estimate the required sample size based on the c.i. half widths w_c and w_u. In this section, we show that it may be possible to estimate the required sample sizes based on w_{cu}, i.e., the c.i. half-width of $\mu_c - \mu_u$.

Assume that w_{cu} is determined by a vector of unknown parameters π and an unknown function Ψ, e.g., $w_{cu} = \Psi(h, n_c, n_u, \pi)$. Then it is necessary to allocate the sample sizes n_c and n_u so that $\Psi(h, n_c, n_u, \pi) \leq d^*$; where d^* is the absolute indifference amount. Unfortunately, the function Ψ is generally unknown.

In the special case that $f(x)$ is the pdf of the t-distribution with $n_0 - 1$ d.f., $\Psi(h, n_c, n_u, \pi)$ is known. Recall that $\vartheta = T_c - T_u$. Then the cdf of ϑ

$$
\begin{aligned}
G(h) &= \Pr[\vartheta \leq h] \\
&= \frac{v!}{(c-1)!(v-c)!} \frac{(k-v)!}{(m-c)!(k-v-m+c-1)!} \times \\
&\quad \int_{-\infty}^{\infty} \int_{-\infty}^{y+h} [F(x)]^{c-1}[1-F(x)]^{v-c} f(x) \times \\
&\quad [F(y)]^{m-c}[1-F(y)]^{k-v-m+c-1} f(y) dx dy.
\end{aligned}
$$

Note that the cdf $G(h)$ is determined only by the d.f. of the t-distribution given c, v, m, k. By definition h is the P^* quantile of the distribution of ϑ when $G(h) = P^*$.

Furthermore,

$$\vartheta = T_c - T_u = \frac{(\tilde{X}_c - \tilde{X}_u) - (\mu_c - \mu_u)}{d^*/h} = \frac{(\tilde{X}_c - \tilde{X}_u) - (\mu_c - \mu_u)}{\sqrt{S_c^2(n_0)/(2n_c) + S_u^2(n_0)/(2n_u)}}.$$

Without loss of generality, we temporarily assume n_c and n_u are real numbers. The last equality holds since

$$\frac{S_c^2(n_0)}{2n_c} = \frac{S_u^2(n_0)}{2n_u} = \frac{(d^*)^2}{2h^2} \text{ and } \sqrt{\frac{S_c^2(n_0)}{2n_c} + \frac{S_u^2(n_0)}{2n_u}} = \frac{d^*}{h}.$$

Consequently,

$$\Pr\left[\frac{(\tilde{X}_c - \tilde{X}_u) - (\mu_c - \mu_u)}{\sqrt{S_c^2(n_0)/(2n_c) + S_u^2(n_0)/(2n_u)}} \le h\right] = P^*.$$

Let

$$w_{cu} = \frac{h}{\sqrt{2}}\sqrt{\frac{S_c^2(n_0)}{n_c} + \frac{S_u^2(n_0)}{n_u}}.$$

Then,

$$\Pr[\tilde{X}_c - \tilde{X}_u - w_{cu} \le \mu_c - \mu_u] = P^*.$$

Hence, to obtain $w_{cu} \le d^*$, $n_i \ge (hS_i(n_0)/d^*)^2$, provided $(hS_i(n_0)/d^*)^2 > n_0$. This is the same result as that of [Dudewicz and Dalal (1975)], where h_1 is obtained with $c = v = m = 1$.

10.1.5 *Adjustment of the Difference of Sample Means*

It is known that (absolute) indifference-zone procedures that are derived from the LFC are conservative and allocate samples based entirely on the variances. The efficiency of the procedures can be improved by taking into account true/sample means. Let $d_i = \mu_i - \mu_{i_1}$ for $i = 1, 2, \ldots, k$. Then $\max(d^*, d_i)$ instead of d^* will be used to compute the required sample sizes. In reality we use $\hat{d}_i$, an estimator of d_i, to compute sample sizes. We involuntarily introduce error into the process, hence, the procedure does not guarantee $\mathrm{P(CS)} \ge P^*$. Nevertheless, we can use a conservative adjustment to increase P(CS). The adjustment takes into consideration the randomness of $\bar{X}_b(n_0)$ and allocates more replications or batches to more promising designs.

Let the adjustment $a = t_{n_0-1,P^*} S_b(\bar{X}_b(n_0))$, where $S_b^2(\bar{X}_b(n_0))$ is the variance of $\bar{X}_b(n_0)$. We use the one-tailed upper P^* confidence limit of μ_b, $U(\bar{X}_b(n_0)) = \bar{X}_b(n_0) + a$, as an estimator of the reference point. If the procedure is to find the maximum, then the one-tailed lower P^* confidence limit should be used. Even though this is somewhat arbitrary, we know that $\Pr[\mu_b \le U(\bar{X}_b(n_0))] \approx P^*$. In this setting,

$$d'_i = \max(d^*, \bar{X}_i(n_0) - U(\bar{X}_b(n_0))). \tag{10.1}$$

Conservative users can use higher confidence of $U(\bar{X}_b(n_0))$ to increase P(CS). We then compute the sample sizes for each design based on the following formula.

$$N_i = \max(n_0 + 1, \lceil (hS_i(n_0)/d'_i)^2 \rceil), \text{ for } i = 1, 2, \ldots, k. \tag{10.2}$$

Let $L(\bar{X}_i(n_0) - \bar{X}_b(n_0))$ be the lower confidence limit of $\bar{X}_i(n_0) - \bar{X}_b(n_0)$. One can also use $d'_i = \max(d^*, L(\bar{X}_i(n_0) - \bar{X}_b(n_0)))$, which requires more computations.

With this adjustment, we will allocate more simulation replications or batches to more promising designs. Let

$$\hat{d}_i = \max(d^*, \bar{X}_i(n_0) - \bar{X}_b(n_0)) \tag{10.3}$$

and

$$N'_i = \max(n_0 + 1, \lceil (hS_i(n_0)/\hat{d}_i)^2 \rceil), \text{ for } i = 1, 2, \ldots, k.$$

If $\bar{X}_i(n_0) = \bar{X}_b(n_0) + \delta$ and N_i and $N'_i > n_0 + 1$, then

$$\frac{N_i}{N'_i} = \begin{cases} 1 & 0 < \delta \le d^* \\ (\delta/d^*)^2 & d^* < \delta \le d^* + a \\ (\delta/(\delta - a))^2 & d^* + a < \delta. \end{cases}$$

Therefore, design i with the first stage sample mean $\bar{X}_i(n_0)$, such that $d^* < \bar{X}_i(n_0) - \bar{X}_b(n_0) \le d^* + a$, will have significantly increased sample sizes with this adjustment. The sample size for that particular design is increased by $(\delta/d^*)^2 - 1$ times. On the other hand, if $d^* + a < \bar{X}_i(n_0) - \bar{X}_b(n_0)$, the additional sample sizes allocated with this adjustment is very minimal. If $\bar{X}_i(n_0) - \bar{X}_b(n_0) \le d^*$, then there are no changes in sample sizes.

We would like to point out that the purpose of R&S procedures is to select a good design i and is not to estimate μ_{i_1}. Because the procedure uses $\bar{X}_b(n_0)$ as an estimator of the reference point, we would like to have some confidence of $\bar{X}_b(n_0)$. If the variance of the samples of the best alternative at the first stage is small, we are more confident with this mean estimator, therefore, the adjustment will be small. In contrast, if the variance is large, we are less confident with this mean estimator, consequently, the adjustment will be large.

10.1.6 *The Source Code of Computing Additional Sample Size*

This subroutine implements the algorithm of computing the additional sample size of each design for next iteration of simulation as discussed in previous section.

```
void etss(double* s_mean,double* s_var,int nd, int* n,float h,
float d,int adj,int *add_budget,int *an)
/* s_mean[i]: sample mean of design i, i=0,1,..,nd-1
   s_var[i]: sample variance of design i, i=0,1,..,nd-1
   nd: the number of designs
   n[i]: number of simulation replication of design i,
         i=0,1,..,nd-1
   h: the critical value,
   d: the indifference amount,
   adj: switch for the adjustment,
   add_budget: the simulation budget, set to -1 if no limit
   an[i]: additional number of simulation replication
          assigned to design i, i=0,1,..,nd-1               */
{
int i;
int b;
double viz;
double adjAmt = 0.0;

      b=best(s_mean, nd);

      if (adj) adjAmt = sqrt(s_var[b]/n[b]);

      for(i=0;i<nd;i++) {
         if (*add_budget == 0 ) {
            an[i] = 0;
            continue;
         };

         viz = s_mean[i] - s_mean[b] - adjAmt;
         if (viz < d) viz = d;
```

```
        an[i] = (ceil(s_var[i]*pow(h/viz,2)) - n[i]+1) / 2;
        if (an[i] < 0) an[i] = 0;

        if (*add_budget > 0) {
           if (*add_budget < an[i])
              an[i] = *add_budget;

           *add_budget -= an[i];

        }; /* if add_budget */
     }; /* for i */
}


int best(float* t_s_mean,int nd)
/* Determines the best design based on current simulation
   results
   t_s_mean[i]: tempary array for sample mean of design i,
                i=0,1,..,ND-1
   nd: the number of designs                                */
{
int i, min_index;
   min_index=0;
   for (i=1;i<nd;i++)
       if(t_s_mean[i]<t_s_mean[min_index])
          min_index=i;

   return min_index;
}
```

10.1.7 *A Sequential Ranking and Selection Procedure (SRS)*

From the discussion in Section 10.1.1, if $w_{cu} \leq \mu_u - \mu_c$, then $\mathrm{P(CS)} \geq P^*$. Of course, the values of μ_u and μ_c are unknown and can not be used to compute the c.i. half-width. In practice, sample means are necessary in order to estimate the required w_{cu}. Sort the sample means such that $\bar{X}_{b_1} \leq \bar{X}_{b_2} \leq \cdots \leq \bar{X}_{b_k}$. Let $U(\bar{X}_{b_v})$ and $L(\bar{X}_{b_{m+1}})$, respectively, be the upper and lower P^* confidence limits of μ_{b_v} and $\mu_{b_{m+1}}$. [Chen and Kelton

(2005)] replace d^* by

$$d_{b_l} = \begin{cases} \max(d^*, L(\bar{X}_{b_{m+1}}) - \bar{X}_{b_l}) & 1 \leq l \leq v \\ \max(d^*, \bar{X}_{b_l} - U(\bar{X}_{b_v})) & v+1 \leq l \leq k \end{cases}$$

when computing the required sample sizes.

We now present cost-effective sequential approach to select a subset of size m that contains c of the v best system from k alternatives. We denote this procedure SRS.

(1) Initialize the set I to include all k designs. Simulate n_0 replications or batches for each design $i \in I$. Set the iteration number $r = 0$, and $N_{1,r} = N_{2,r} = \ldots = N_{k,r} = n_0$, where $N_{i,r}$ is the sample size allocated for design i at the rth iteration.
(2) Set $r = r+1$ and compute $\delta_{i,r}$, the incremental number of replications or batches for design i at the rth iteration according to equation Eq. (10.4).
(3) If $\delta_{i,r} = 0$ and $i \neq b$ (where $\bar{X}_{b,r} = \min_{i \in I} \bar{X}_{i,r}$), delete design i from the subset I.
(4) If there is only one element in the subset I, go to step 6.
(5) Simulate $\delta_{i,r}$ additional replications or batches for each design $i \in I$ at the rth iteration. Go to step 2.
(6) Calculate and rank the sample means such that $\bar{X}_{b_1} \leq \bar{X}_{b_2} \leq \ldots \leq \bar{X}_{b_k}$. Select system b_l iff $\bar{X}_{b_l} \leq \bar{X}_{b_m}$.

Let $d_{i,r}$ denote d_i at the rth iteration. The additional sample size for alternative i at iteration $r+1$ is

$$\delta_{i,r+1} = \lceil ((hS_i(N_{i,r})/d_{i,r})^2 - N_{i,r})^+/2 \rceil, \text{ for } i = 1, 2, \ldots, k. \quad (10.4)$$

Here $(x)^+ = \max(0, x)$ and the critical constant h depends on (c, v, m, k, P^*, n_0). Note that $h(1, 1, 1, k, P^*, n_0) = h_1(k, P^*, n_0)$. Furthermore,

$$N_{i,r+1} = N_{i,r} + \delta_{i,r+1}.$$

Let us consider the steps between taking additional samples, that is steps 2 through 5, be one iteration. We can reduce the number of iterations with a larger incremental sample size for system i at the rth iteration, but we run the risk of allocating more samples than necessary to non-promising systems. We believe use Eq. (10.4) to compute the additional sample size for alternative i at iteration $r+1$ is a good compromise.

Note that sample means $\bar{X}_i$, instead of weighted sample means $\tilde{X}_i$, are used to determine the subset. This is because there are more than two stages of sample means and we can no longer use the approach of [Dudewicz and Dalal (1975)] to compute weighted sample means. While $\bar{X}_i$ for $i = 1, 2, \ldots, k$ are still t-distributed, they have different degrees of freedom. This sequential procedure is asymptotically valid and performs well in terms of P(CS) and sample sizes. For a discussion of the performance of using the weighted sample means and the overall sample means in two-stage procedures, please see [Chen (2011)].

10.2 Some Extensions of Selection of Continuous Distributions

This section discusses several variations of selection procedures: restricted subset selection, optimal subset selection, relative indifference-zone selection, and multiple comparison with the best.

10.2.1 *Restricted Subset Selection*

This section investigates the problem of restricted subset selection, i.e., the selected subset attempts to exclude systems that are deviated more than d^* from the best. It can be shown that the sample sizes should be large enough such that the c.i. half width $w_{cu} \leq d^*/2$. In other words, the required sample size for system i is $n_i \geq (2hS_i(n_0)/d^*)^2$, provided $(2hS_i(n_0)/d^*)^2 > n_0$. The procedure selects system b_l if and only if $\tilde{X}_{b_l} \leq \tilde{X}_{b_1} + d^*/2$. The size of the selected subset is random but at most m populations (which is specified by users) will finally be chosen.

In the special case that $\mu_{i_1} = \mu_{i_2} = \cdots = \mu_{i_k}$, we want to select all good systems. We can write

$$\begin{aligned}
\mathrm{P(CS)} &= \Pr[\tilde{X}_{b_k} < \tilde{X}_{b_1} + d^*/2] \\
&= \Pr\left[\frac{\tilde{X}_{b_k} - \mu_{b_k}}{d^*/(2h_3)} < \frac{\tilde{X}_{b_1} - \mu_{b_1}}{d^*/(2h_3)} - \frac{\mu_{b_k} - \mu_{b_1} - d^*/2}{d^*/(2h_3)}\right] \\
&= \Pr\left[T_{b_k} < T_{b_1} + h_3\right] \\
&= \int_{-\infty}^{\infty} G_{k:k}(t_{b_1} + h_3) dG_{1:k}(t_{b_1}).
\end{aligned}$$

The right-hand side is then equated to P^* to solve for h_3. The required sample size for system i is then $n_i \geq (2h_3S_i(n_0)/d^*)^2$, provided $(2h_3S_i(n_0)/d^*)^2 > n_0$.

In the special case that $\mu_{i_1} + d^* = \mu_{i_2} = \cdots = \mu_{i_k}$, we want to select only the best system. Let $\tilde{X}_{c_2} = \min_{l=2}^{k} \mu_{i_l}$. We can write

$$\begin{aligned}
\text{P(CS)} &= \Pr[\tilde{X}_{c_2} > \tilde{X}_{i_1} + d^*/2] \\
&= \Pr\left[\frac{\tilde{X}_{c_2} - \mu_{i_2}}{d^*/(2h_1)} > \frac{\tilde{X}_{i_1} - \mu_{i_1}}{d^*/(2h_1)} - \frac{\mu_{c_2} - \mu_{i_1} - d^*/2}{d^*/(2h_1)}\right] \\
&= \Pr\left[T_{i_1} < T_{c_2} + h_1\right] \\
&= \int_{-\infty}^{\infty} F(t_{c_2} + h_1) dG_{1:k-1}(t_{c_2}).
\end{aligned}$$

Since $h_3 > h_1$, the sample sizes that guarantee all the sample means of the best systems are within $d^*/2$ from the best sample mean (with the specified probability P^*) also guarantee the sample means of non-d^*-near-best systems (i.e., systems with mean greater than $\mu_{i_1} + d^*$) will not be within $d^*/2$ from the best sample means (with no less the specified probability P^*).

A Sequential Restricted Subset Selection (SRSS) Procedure

(1) Let $N_{i,r}$ be the sample size allocated for system i and $\bar{X}_{i,r}$ be the sample mean of system i at the rth iteration. Simulate n_0 samples for all systems. Set the iteration number $r = 0$, and $N_{1,r} = N_{2,r} = \cdots = N_{k,r} = n_0$. Note that for $r \geq 1$, $N_{i,r}$ can have different values for different i. Specify the value of the indifference amount d^* and the required precision P^*.

(2) Calculate the sample means and sample variances. Rank the sample means such that $\bar{X}_{b_1} \leq \bar{X}_{b_2} \leq \ldots \leq \bar{X}_{b_k}$.

(3) Calculate the required sample size $N_{b_l,r+1} = \max(n_0, \lceil (2h_3 S_{b_l}(N_{b_l,r})/d_{b_l})^2 \rceil)$, for $l = 1, 2, \ldots, k$. Here,

$$d_{b_l} = \begin{cases} \max(d^*, L(\bar{X}_{b_{m+1}}) - \bar{X}_{b_l}) & 1 \leq l \leq v \\ \max(d^*, \bar{X}_{b_l} - U(\bar{X}_{b_v})) & v+1 \leq l \leq k. \end{cases}$$

(4) If $N_{i,r+1} \leq N_{i,r}$, for $i = 1, 2, \ldots, k$, go to step 6.

(5) Simulate additional $\lceil (N_{i,r+1} - N_{i,r})^+/2 \rceil$ samples for system i. Set $r = r + 1$. Go to step 2.

(6) Select system b_l iff $\bar{X}_{b_l} \leq \min(\bar{X}_{b_m}, \bar{X}_{b_1} + d^*/2)$.

10.2.2 *An Indifference-Zone Procedure to Select Only and/or All The Best Systems*

In this section, we develop an indifference-zone procedure to select the optimal subset, i.e., the subset contains only and/or all the best systems. We use the parameter δ to guide our decision. Based on the restricted-subset-selection procedure, we consider the case that the indifference amount $d^* = 2\delta$. That is, we will set $\delta = d^*/2$ when the indifference amount d^* is specified. For a specified confidence level P^*, the subset is guaranteed to contain all the best systems and none of the systems that deviate more than d^* from the best system(s). The size of the subset is determined by the procedure and users do not need to specify the upper bound.

The procedure satisfies the following:

(1) If $\mu_{i_1} = \mu_{i_2} = \cdots = \mu_{i_k}$, then
$$\Pr[\tilde{X}_{b_k} \le \tilde{X}_{b_1} + \delta] \ge P^*.$$
(2) If $\mu_{i_1} + 2\delta = \mu_{i_2} = \cdots = \mu_{i_k}$, then
$$\Pr[\tilde{X}_{c_2} > \tilde{X}_{i_1} + \delta] \ge P^*.$$
Here $\tilde{X}_{c_2} = \min_{l=2}^{k} \tilde{X}_{i_l}$.

(3) In the cases that $\mu_{i_1} = \mu_{i_2} = \cdots = \mu_{i_v}$ and $\mu_{i_v} + 2\delta = \mu_{i_{v+1}} = \cdots = \mu_{i_k}$, where $2 \le v \le k-1$. Let $\tilde{X}_{c_1}$ and $\tilde{X}_{c_v}$ be the smallest and largest weighted sample mean of systems i_l for $l = 1, 2, \ldots, v$, respectively. Let $\tilde{X}_{c_u}$ be the smallest weighted sample mean of systems i_l for $l = v+1, v+2, \ldots, k$. We have
$$\Pr[\tilde{X}_{c_v} \le \tilde{X}_{c_1} + \delta] \ge P^*$$
and
$$\Pr[\tilde{X}_{c_u} > \tilde{X}_{c_1} + \delta] \ge P^*.$$

Recall that $\tilde{X}_{b_1} = \min_{l=1}^{k} \tilde{X}_{i_l}$, $\tilde{X}_{b_k} = \max_{l=1}^{k} \tilde{X}_{i_l}$, $\tilde{X}_{c_1} = \min_{l=1}^{v} \tilde{X}_{i_l}$, and $\tilde{X}_{c_v} = \max_{l=1}^{v} \tilde{X}_{i_l}$. Because $\tilde{X}_{c_v} \sim G_{v:v}(t)$ and $\tilde{X}_{c_1} \sim G_{1:v}(t)$ while $\tilde{X}_{b_k} \sim G_{k:k}(t)$ and $\tilde{X}_{b_1} \sim G_{1:k}(t)$ and $v < k$, $\Pr[\tilde{X}_{c_v} \le \tilde{X}_{c_1} + \delta] \ge \Pr[\tilde{X}_{b_k} \le \tilde{X}_{b_1} + \delta]$. Note that $\mathrm{E}(\tilde{X}_{c_v}) \le \mathrm{E}(\tilde{X}_{b_k})$, $\mathrm{E}(\tilde{X}_{b_1}) \le \mathrm{E}(\tilde{X}_{c_1})$, $\mathrm{Var}(\tilde{X}_{c_v}) \ge \mathrm{Var}(\tilde{X}_{b_k})$, and $\mathrm{Var}(\tilde{X}_{c_1}) \ge \mathrm{Var}(\tilde{X}_{b_1})$. Figure 10.2 shows the empirical distributions of the first order statistic of $k = 1, 5$, and 9 $N(0,1)$ random variables. Hence, the sample sizes that guarantee case 1 (i.e. $\Pr[\tilde{X}_{b_k} \le \tilde{X}_{b_1} + \delta] \ge P^*$) also guarantee $\Pr[\tilde{X}_{c_v} \le \tilde{X}_{c_1} + \delta] \ge P^*$. Recall that $\tilde{X}_{c_2} = \min_{l=2}^{k} \tilde{X}_{i_l}$ and $\tilde{X}_{c_u} = \min_{l=v+1}^{k} \tilde{X}_{i_l}$. Because $\tilde{X}_{c_u} \sim G_{1:k-v}(t)$ and $\tilde{X}_{c_1} \sim G_{1:v}(t)$ while $\tilde{X}_{c_2} \sim G_{1:k-1}(t)$ and $k - v \le k - 1$, $\Pr[\tilde{X}_{c_u} \ge \tilde{X}_{c_1} + \delta] \ge \Pr[\tilde{X}_{c_2} \ge \tilde{X}_{i_1} + \delta]$. Furthermore, $\mathrm{E}(\tilde{X}_{c_u}) \ge \mathrm{E}(\tilde{X}_{c_2})$, $\mathrm{E}(\tilde{X}_{i_1}) \ge \mathrm{E}(\tilde{X}_{c_1})$, $\mathrm{Var}(\tilde{X}_{c_u}) \ge \mathrm{Var}(\tilde{X}_{c_2})$, and $\mathrm{Var}(\tilde{X}_{i_1}) \ge \mathrm{Var}(\tilde{X}_{c_1})$. Hence, the sample sizes that guarantee case 2 (i.e. $\Pr[\tilde{X}_{c_2} > \tilde{X}_{i_1} + \delta] \ge P^*$) also guarantee $\Pr[\tilde{X}_{c_u} > \tilde{X}_{c_1} + \delta] \ge P^*$. Consequently, as long as the first

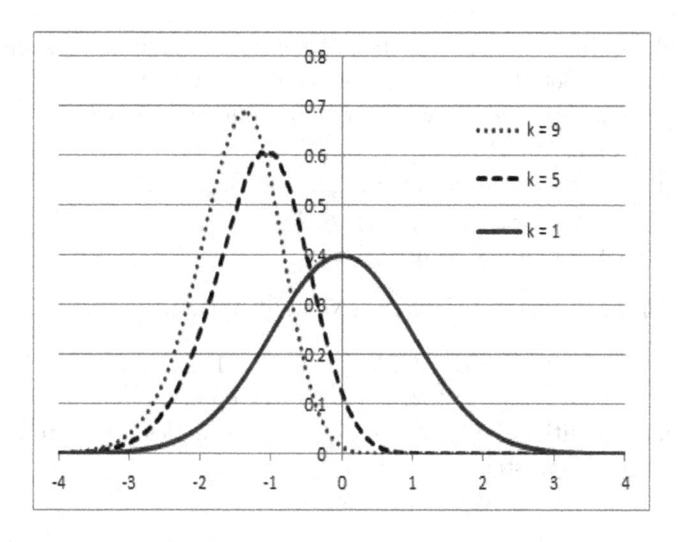

Fig. 10.2 First order statistic of k $N(0,1)$ random variables

two cases are satisfied, case 3 will be satisfied as well. Systems l with $\mu_{i_1} < \mu_{i_l} \leq \mu_{i_1} + 2\delta$ may be included in the final subset, but there is no probability guarantees. Furthermore, the probability of systems l being included in the final subset decreases as μ_{i_l} deviates farther away from μ_{i_1}.

For case 1, the probability guarantee is achieved when the sample size for system i n_{ci} is computed by Eq. (4.4). We now derive the required sample size to achieve the probability guarantee for case 2. We can write

$$\begin{aligned} \mathrm{P(CD)} &= \Pr[\tilde{X}_{c_2} > \tilde{X}_{i_1} + \delta] \\ &= \Pr[T_{i_1} < T_{c_2} + h_1] \\ &= \int_{-\infty}^{\infty} F(t + h_1) g_{1:k-1}(t) dt. \end{aligned}$$

We equate the right-hand side to P^* and solve for h_1. The required sample size for system i n_{di} for system i is then computed by Eq. (4.2) with d^* replaced by δ. Note that the h_1 values are available from the tables in [Law (2014)]. It is known that in this setting $n_{di} \leq n_{ci}$, hence, $n_i = \max(n_{ci}, n_{di}) = n_{ci}$.

We now present a cost-effective sequential approach to select the optimal subset. We denote this approach the SOSS (Sequential Optimal Subset Selection) procedure.

(1) Let $N_{i,r}$ be the sample size allocated for system i and $\bar{X}_{i,r}$ be the

sample mean of system i at the rth iteration. Simulate n_0 samples for all systems. Set the iteration number $r = 0$, and $N_{1,r} = N_{2,r} = \cdots = N_{k,r} = n_0$. Specify the value of the indifference amount d^* and the required precision P^*.

(2) Calculate the sample means and sample variances. Rank the sample means such that $\bar{X}_{b_1} \leq \bar{X}_{b_2} \leq \ldots \leq \bar{X}_{b_k}$.

(3) Calculate the required sample size

$$N_{b_l,r+1} = \max(n_0, \lceil (2h_3 S_{b_l}(N_{b_l,r})/d_{b_l})^2 \rceil), \text{ for } l = 1, 2, \ldots, k.$$

Here, $d_{b_l} = \max(d^*, \bar{X}_{b_l} - U(\bar{X}_{b_1}))$, for $l = 1, 2, \ldots, k$.

(4) If $N_{i,r+1} \leq N_{i,r}$, for $i = 1, 2, \ldots, k$, go to step 6.

(5) Simulate additional $\lceil (N_{i,r+1} - N_{i,r})^+/2 \rceil$ samples for system i. Set $r = r + 1$. Go to step 2.

(6) Select system b_l iff $\bar{X}_{b_l} \leq \bar{X}_{b_1} + d^*/2$.

The critical value h_3 depends on k, n_0, and P^*. Even though the sample sizes for each system change at each iteration, we use the initial value of h_3 through all iterations. This simplifies programming efforts and provides conservative estimates of the sample sizes.

10.2.3 *Ratio Statistics of Variance of Normally Distributed Variables*

The range statistics and/or the absolute indifference zone is applied when the parameter of interest is a location parameter. On the other hand, the ratio statistics and/or the relative indifference zone needs to be used when the parameter of interest is a scale parameter, e.g., variance. The variance of X_{ij} quantifies the dispersion of X_{ij} and is denoted by $\text{Var}(X_{ij}) = \sigma_i^2$ and, if it exists, $\sigma_i^2 = E[(X_{ij} - \mu_i)^2]$, where $\mu_i = E(X_{ij})$ is the mean of X_{ij}. Let n_i denote the number of samples of the ith system. The sample variance with n_i observations $S_i^2(n_i) = \sum_{j=1}^{n_i}(X_{ij} - \bar{X}_i(n_i))^2/(n_i - 1)$ is the unbiased variance estimator of σ_i^2 and $\bar{X}_i(n_i) = \sum_{j=1}^{n_i} X_{ij}/n_i$ is the unbiased mean estimator of μ_i. When X_{ij}'s are normally distributed, the variables $\chi_i = (n_i - 1)S_i^2(n_i)/\sigma_i^2$ for $i = 1, 2, \ldots, k$ are independent χ^2 random variables with $n_i - 1$ degrees of freedom. It is known that $E(\chi_i) = n_i - 1$ and $\text{Var}(\chi_i) = 2(n_i - 1)$. Hence, $E(S_i^2(n_i)) = \sigma_i^2$ and $\text{Var}(S_i^2(n_i)) = 2\sigma_i^4/(n_i - 1)$.

Let n_e be the sample size of all k systems, i.e., $n_i = n_e$ for $i = 1, 2, \ldots, k$. We use the notation S_i^2 instead of $S_i^2(n_i)$ for the rest of this section. Let

χ_{b_l} be the lth smallest of χ_i such that $\chi_{b_1} \leq \chi_{b_2} \leq \cdots \leq \chi_{b_k}$. Because the distribution of χ_i for $i = 1, 2, \ldots, k$, is known, given the sample size (i.e., the d.f.), the quantile of the distribution of χ_{b_k}/χ_{b_1} can be determined. For example, let δ_r be the 0.90 quantile of the distribution of χ_{b_2}/χ_{b_1} when $k = 2$, then $\Pr[\chi_{b_2}/\chi_{b_1} \leq \delta_r] = 0.90$. Hence, we do not reject the null hypothesis that $\sigma_1^2 = \sigma_2^2$ with 0.90 confidence, when $\chi_{b_2}/\chi_{b_1} \leq \delta_r$, or similarly, $S_{b_2}^2/S_{b_1}^2 \leq \delta_r$. Furthermore, given a user specified δ_r, the required sample size n_e to achieve $\Pr[\chi_{b_2}/\chi_{b_1} \leq \delta_r] = 0.90$ can be determined. In the cases that $k > 2$, the quantile of the distribution of χ_{b_k}/χ_{b_1} can be used to test the null hypothesis that $\sigma_1^2 = \sigma_2^2 = \cdots = \sigma_k^2$.

We now assess P(CS) of this selection problem, i.e., to select the optimal subset of systems having the smallest variances from k alternatives. Let $S_{b_1}^2$ and $S_{b_k}^2$, respectively, be the smallest and the largest sample variances from σ_i^2 for $i = 1, 2, \ldots, k$ and let $\sigma_{b_1}^2$ and $\sigma_{b_k}^2$, respectively, be their unknown true variances. That is, $S_{b_1}^2$ and $S_{b_k}^2$ are, respectively, the first and the kth order statistics. Note that systems b_1 and b_k are likely to be different in different replications. Furthermore, let $\delta_r > 1$ be a user specified upper bound of the ratio for the hypothesis test. System i having $S_i^2/S_{b_1}^2 < \delta_r$ will be considered feasible, i.e., there is no significant difference between σ_i^2 and $\sigma_{b_1}^2$. Then, under the (equivalence) null hypothesis that $\sigma_1^2 = \sigma_2^2 = \cdots = \sigma_k^2$ (and consequently, $\sigma_{b_1}^2 = \sigma_{b_k}^2$),

$$\begin{aligned}
\mathrm{P(CS)} &= \Pr[S_{b_k}^2/S_{b_1}^2 < \delta_r] \\
&= \Pr\left[\frac{(n_e-1)S_{b_k}^2/\sigma_{b_k}^2}{(n_e-1)S_{b_1}^2/\sigma_{b_1}^2}\frac{\sigma_{b_k}^2}{\sigma_{b_1}^2} < \delta_r\right] \\
&= \Pr\left[\chi_{b_k}/\chi_{b_1} < \delta_r\right].
\end{aligned}$$

The third equality follows because $\chi_{b_k} = (n_e - 1)S_{b_k}^2/\sigma_{b_k}^2$, and $\chi_{b_1} = (n_e - 1)S_{b_1}^2/\sigma_{b_1}^2$. Note that χ_{b_1} and χ_{b_k} are, respectively, the first and the kth order statistics of $(n_e - 1)S_i^2/\sigma_i^2$ for $i = 1, 2, \ldots, k$. Let ω and Ω, respectively, denote the pdf and cdf of the χ^2 distribution with $n_e - 1$ d.f. Then, $\chi_{b_1} \sim G_{1:k}(\chi)$ and $\chi_{b_k} \sim G_{k:k}(\chi)$ with f and F, respectively, replaced by ω and Ω. Figure 10.3 shows the empirical distributions of the first order statistic of $k = 1, 5$, and 9 χ^2 (with 19 d.f.) random variables.

Because χ_{b_1} and χ_{b_k} are no longer χ^2 distributed when $k \geq 2$, the random variable χ_{b_k}/χ_{b_1} is no longer F-distributed. Nevertheless,

$$\mathrm{P(CS)} \geq \int_0^\infty G_{k:k}(\chi_{b_1}\delta_r)g_{1:k}(\chi_{b_1})d\chi_{b_1}.$$

Let j_l be the system having the lth smallest variance such that $\sigma_{j_1}^2 \leq \sigma_{j_2}^2 \leq \cdots \leq \sigma_{j_k}^2$. Let $d_r^* > 1$ be the relative indifference amount, i.e., we will not

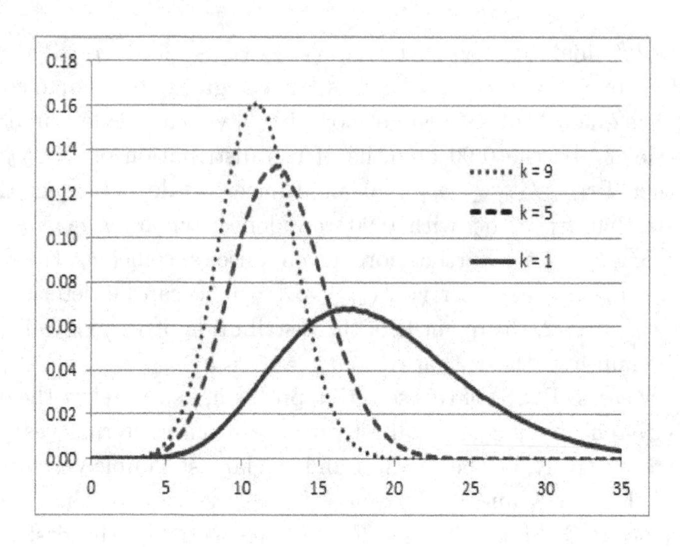

Fig. 10.3 First order statistic of k χ^2 (19 d.f.) random variables

distinguish system j_1 and systems i having $\sigma_i^2 < \sigma_{j_1}^2 d_r^*$. The LFC (least favorable configuration) of relative indifference zone is $\sigma_{j_1}^2 d_r^* = \sigma_{j_2}^2 = \cdots = \sigma_{j_k}^2$.

Given the user specified δ_r, we equate the right-hand side to P^* to determine the required d.f. of the underlying χ^2 distribution Ω, i.e., the required sample size n_e. To perform the null hypothesis test that all k variances are equal, we obtain the required sample size n_e given the variances ratio δ_r. Simulate n_e samples for each system and compute the sample variances S_i^2 for $i = 1, 2, \ldots, k$. We do not reject the null hypothesis that $\sigma_1^2 = \sigma_2^2 = \cdots = \sigma_k^2$ when $S_{b_k}^2/S_{b_1}^2 < \delta_r$.

For the un-equivalence cases, under the LFC, we would like to ensure that system j_l for $l = 2, 3, \ldots, k$ will not be accepted. Let $S_{c_2}^2 = \min_{l=2}^k S_{j_l}^2$ and let n_f denote the required sample size of each system. We can write

$$\begin{aligned}
\mathrm{P(CS)} &= \Pr[S_{c_2}^2/S_{j_1}^2 \geq \delta_r] \\
&= \Pr\left[\frac{(n_f-1)S_{c_2}^2/\sigma_{c_2}^2}{(n_f-1)S_{j_1}^2/\sigma_{j_1}^2}\frac{\sigma_{c_2}^2}{\sigma_{j_1}^2} \geq \delta_r\right] \\
&= \Pr\left[\chi_{c_2} \geq \chi_{j_1}\delta_r/d_r^*\right] \\
&= \Pr\left[\chi_{j_1}/\chi_{c_2} \leq d_r^*/\delta_r\right].
\end{aligned}$$

The third equality follows because $\chi_{c_2} = (n_f-1)S_{c_2}^2/\sigma_{c_2}^2$ and $\chi_{j_1} = (n_f - 1)S_{j_1}^2/\sigma_{j_1}^2$. Note that χ_{c_2} is the first order statistic of $(n_f-1)S_i^2/\sigma_i^2$ for

$i = 2, 3, \ldots, k$. We equate the right-hand side to P^* to determine the required d.f. of the underlying χ^2 distribution, i.e., the required sample size n_f. Note that the sample sizes n_e and n_f can be estimated by the procedure of [Chen and Li (2010)]. When the absolute indifference amount is d^*, we set the amount for comparison with a control $\delta = d^*/2$. When the relative indifference amount is d_r^*, we set the amount for comparison with a control $\delta_r = \sqrt{d_r^*}$. In this setting,

$$\mathrm{P(CS)} \geq \int_0^\infty \Omega(\chi_{c_2}\delta_r) g_{1:k-1}(\chi_{c_2}) d\chi_{c_2}.$$

The underlying pdf (ω) and cdf (Ω) of the order-statistic distribution $g_{1:k-1}(x)$ are the χ^2 distribution with $n_f - 1$ d.f. Consequently, if we compute the sample variances with $n = \max(n_e, n_f)$ samples and select systems i having $S_i^2/S_{b_1}^2 < \delta_r$, then the selected subset contains only and/or all the best systems with probability no less than P^*. Again, good systems i having $\sigma_i^2 < \sigma_{j_1}^2 d_r^*$ may be included in the selected subset, however, there is no probability guarantees. See [Chen (2008)] for an example of selecting a subset up to size m contains at least c of the v systems having the smallest variances from k alternatives.

In cases of the relative indifference zone, the single-stage procedure allocates sample sizes equally among systems with the fixed total sample size $N = kn_0$. Under the non-LFC, allocating sample sizes equally among systems is not optimal. Let Y_i be the estimate of the performance measure of system i and $Y_{b_1} \leq Y_{b_2} \leq \cdots \leq Y_{b_k}$. [Chen (2008)] proposes allocating sample sizes according to the following rules to increase P(CS):

$$\frac{n_{b_i}}{n_{b_j}} = \begin{cases} \frac{Y_{b_i}}{Y_{b_j}} & 1 \leq i, j \leq v, \\ 1 & v \leq i, j \leq v+1, \\ \frac{Y_{b_j}}{Y_{b_i}} & v+1 \leq i, j \leq k. \end{cases} \tag{10.5}$$

The two-stage selection procedure proceeds as follows:

(1) Specify the maximum total sample size N and the initial sample size $n_0 < N/k$.
(2) Simulate n_0 samples for each system $i = 1, 2, \ldots, k$.
(3) Compute the estimate Y_i with n_0 samples for each system $i = 1, 2, \ldots, k$.
(4) Rank the Y_i and allocate n_i according to the ratios of Eq. (10.5). Note that $\sum_{i=1}^k n_i = N$.
(5) Simulate additional $\max(0, n_i - n_0)$ samples for each system $i = 1, 2, \ldots, k$.

(6) Compute and rank the estimate Y_i with n_i samples for each system $i = 1, 2, \ldots, k$.
(7) Rank the sample variances such that $S^2_{c_1}(n_{c_1}) \leq S^2_{c_2}(n_{c_2}) \leq \cdots \leq S^2_{c_k}(n_{c_k})$. Return the set $\{c_1, c_2, \ldots, c_m\}$.

Note that the ratios of Eq. (10.5) may not be precisely achieved when the specified N is not large enough. This is because the procedure can not take back the computation budget that has already been spent. Furthermore, the two-stage procedure can easily be extended to a sequential procedure. For example, instead of allocating all the additional samples in the second stage, only a portion of additional samples can be allocated and used to recompute the ratios as the procedure proceeds.

10.2.4 *Multiple Comparisons with the Best*

Multiple comparisons provide simultaneous confidence intervals on selected differences among the systems. It is known that indifference-zone selection procedures also guarantee that the c.i. coverage of multiple comparisons with the best (MCB) to have the same confidence level of the selection procedures. These c.i.'s bound the differences between the performance of each system and the best of the others with a pre-specified confidence level. Let w_{ij} be the one-tailed $P = 1 - (1 - P^*)/(k - 1)$ c.i. half-width. These MCB c.i.'s are

$$\Pr[\mu_i - \min_{j \neq i} \mu_j \in [\max_{j \neq i}(\bar{X}_i - \bar{X}_j - w_{ij})^-, \max_{j \neq i}(\bar{X}_i - \bar{X}_j + w_{ij})^+], \forall i] \geq P^*.$$

Here $(x)^-$ denotes $\min(0, x)$ and $(x)^+$ denotes $\max(0, x)$.

Define the events

$$E = \{\mu_i - \mu_{i_1} \leq \bar{X}_i - \bar{X}_{i_1} + w_{ii_1}, \forall i \neq i_1\},$$

$$E_L = \{\mu_i - \min_{j \neq i} \mu_j \geq \max_{j \neq i}(\bar{X}_i - \bar{X}_j - w_{ij})^-, \forall i\},$$

$$E_U = \{\mu_i - \min_{j \neq i} \mu_j \leq \max_{j \neq i}(\bar{X}_i - \bar{X}_j + w_{ij})^+, \forall i\},$$

$$E_T = \{\mu_i - \min_{j \neq i} \mu_j \in [\max_{j \neq i}(\bar{X}_i - \bar{X}_j - w_{ij})^-, \max_{j \neq i}(\bar{X}_i - \bar{X}_j + w_{ij})^+], \forall i\}.$$

Note that E is the event that the upper one-tailed confidence intervals for MCC (Multiple Comparison with a Control), with the control being design i_1, contain all of the true differences $\mu_i - \mu_{i_1}$. Since $\Pr[\mu_i - \mu_{i_1} \leq \bar{X}_i - \bar{X}_{i_1} + w_{ii_1}] \geq P \ \forall i$, $\Pr[E] \geq P^*$. Now following an argument developed by [Edwards and Hsu (1983)], we have that $E \subset E_L \cap E_U$, which will establish the result $\Pr[E_T] \geq P^*$.

First we prove that $E \subset E_L$:

$$\begin{aligned} E &\subset \{\mu_{i_1} - \mu_j \geq \bar{X}_{i_1} - \bar{X}_j - w_{i_1 j}, \forall j \neq i_1\} \\ &\subset \{\mu_{i_1} - \mu_{i_2} \geq \bar{X}_{i_1} - \bar{X}_j - w_{i_1 j}, \forall j \neq i_1\} \\ &\subset \{\mu_i - \mu_{i_2} \geq \max_{j \neq i}(\bar{X}_i - \bar{X}_j - w_{ij})^-, \forall i\} \\ &\subset \{\mu_i - \min_{j \neq i} \mu_j \geq \max_{j \neq i}(\bar{X}_i - \bar{X}_j - w_{ij})^-, \forall i\}, \end{aligned}$$

where the second step follows since $\mu_{i_1} - \mu_{i_2} \geq \mu_{i_1} - \mu_j$ for all $j \neq i_1$ and the third step follows since $\mu_i - \mu_{i_2} \geq 0$ for all $i \neq i_1$ and $(x)^- \leq 0$.

Now we show $E \subset E_U$.

$$\begin{aligned} E &\subset \{\mu_i - \mu_{i_1} \leq \max_{j \neq i}(\bar{X}_i - \bar{X}_j + w_{ij}), \forall i \neq i_1\} \\ &\subset \{\mu_i - \min_{j \neq i} \mu_j \leq \max_{j \neq i}(\bar{X}_i - \bar{X}_j + w_{ij})^+, \forall i\}, \end{aligned}$$

where the first step follows since $\max_{j \neq i}(\bar{X}_i - \bar{X}_j + w_{ij}) \geq \bar{X}_i - \bar{X}_{i_1} + w_{ii_1}$ for all $i \neq i_1$ and the last step follows since $\mu_{i_1} - \min_{j \neq i_1} \mu_j \leq 0$ and $(x)^+ \geq 0$. Hence, $E \subset E_L \cap E_U$, and the proof is complete. See [Edwards and Hsu (1983)] for more detail on multiple comparisons.

Indifference-zone procedures derived based on the LFC achieve $w_{ij} \leq d^*$ when $c = v = m = 1$ and the MCB c.i.'s are simplified to

$$\Pr[\mu_i - \min_{j \neq i} \mu_j \in [(\hat{\mu}_i - \min_{j \neq i} \hat{\mu}_j - d^*)^-, (\hat{\mu}_i - \min_{j \neq i} \hat{\mu}_j + d^*)^+], \forall i] \geq P^*.$$

However, these tight c.i's come at a cost. Our procedure takes into account the differences of sample means, hence, the c.i. half-width w_{ii_1} is around $\max(d^*, \mu_i - \mu_{i_1})$ instead of d^*. Recall that the objective is to select a good design and not to estimate the difference of sample means. This is consistent with the philosophy of ordinal comparison [Ho et al. (1992)] that it is advantageous to rank estimators while the precision of the estimates are still poor when the objective is to select a good design.

10.3 Lognormally Distributed Samples

Let $N(\mu, \sigma^2)$ denote the normal distribution with mean μ and variance σ^2. Let $LN(\mu, \sigma^2)$ denote the lognormal distribution with parameter μ and σ^2. The variable X is $LN(\mu, \sigma^2)$ only if the variable $Y = \ln X$ is $N(\mu, \sigma^2)$. Note that the mean of $LN(\mu, \sigma^2)$ is $\exp(\mu + \sigma^2/2)$. It is clear that unless

σ^2 is known, inferences based on μ alone from the transformed variable Y are not sufficient for the inference on a parameter that is a function of both μ and σ^2.

Consider k independent, lognormally-distributed random variables with parameters μ_i and σ_i^2, $i = 1, 2, \ldots, k$. It is assumed that μ_i and σ_i^2 are unknown and unequal. [John and Chen (2006)] procedure can select the system having the largest linear combination of μ_i and σ_i^2, i.e., the largest $\theta_i(a, b) = \exp(a\mu_i + b\sigma_i^2)$. We are interested in selecting the system with the largest mean, i.e., we limit the discussion to the case that $\theta_i = \theta_i(1, 1/2)$. However, the techniques proposed here can be used to select the largest $\theta_i(a, b)$ for any combinations of a and b as well.

Since θ_i is a scale parameter of the exponential function exp, we apply the relative indifference amount. Let θ_{i_l} be the lth largest mean such that $0 < \theta_{i_1} \leq \theta_{i_2} \leq \ldots \leq \theta_{i_k}$. If the ratio $1 \leq \theta_{i_k}/\theta_{i_{k-1}} < d_r^*$, we say that the systems are in the (relative) indifference zone for correct selection. On the other hand, if the ratio $\theta_{i_k}/\theta_{i_{k-1}} \geq d_r^*$, we say that the systems are in the preference zone for correct selection. Furthermore, the minimal P(CS) (probability of CS) occurs under the LFC, i.e., $\theta_{i_1} = \theta_{i_2} = \ldots = \theta_{i_{k-1}} = \theta_{i_k}/d_r^*$.

Let $Y_{ij} = \ln X_{ij}$. We then compute $\bar{Y}_i$ (sample mean of Y_{ij}) and S_i^2, the unbiased estimator of μ_i and σ_i^2, from all samples. Let $\hat{\beta}_i = \ln \hat{\theta}_i = \bar{Y}_i + S_i^2/2$, and select the systems with the largest $\hat{\beta}_i$ as the one having mean θ_{i_k}. Note that $\hat{\beta}_i$ is strictly increasing of $\hat{\theta}_i$. Furthermore, $\beta_i = \ln \theta_i = \mu_i + \sigma_i^2/2$ and under the LFC $\beta_{i_k} - d_a^* = \beta_{i_l}$ for $l = 1, 2, \ldots, k-1$. That is, the indifference amount d_a^* has been transformed from the relative form (i.e., $\theta_{i_k}/\theta_{i_{k-1}} = d_r^*$) to the absolute form (i.e., $\beta_{i_k} - \beta_{i_{k-1}} = d_a^*$). Note that in this case $d_a^* = \ln d_r^*$.

10.3.1 *The Property of the Constant h_L*

In this section, we define the constant h_L and describe an approach to find its value. Without loss of generality, assume that the systems are under the LFC and system i_k is the best system. Let $S_{i,0}^2$ be the unbiased estimator of σ_i^2 with n_0 observations. For $i = 1, 2, \ldots, k$, let

$$W_i = \frac{\sigma_i^2 + \sigma_i^4/2}{n_i}, \text{ and } w_i = \frac{\sigma_i^2}{S_{i,0}^2} + \frac{\sigma_i^4}{S_{i,0}^4}.$$

For $l = 1, 2, \ldots, k-1$, let

$$Z_{i_l} = \frac{(\hat{\beta}_{i_l} - \hat{\beta}_{i_k}) + d_a^*}{\sqrt{W_{i_l} + W_{i_k}}},$$

$$D_{i_l} = \frac{d_a^*}{\sqrt{W_{i_l} + W_{i_k}}},$$

and

$$Q_{i_l} = \frac{h_L}{\sqrt{w_{i_l} + w_{i_k}}}.$$

We can write the probability of correct selection as:

$$\begin{aligned}
\mathrm{P(CS)} &= \Pr[\hat{\beta}_{i_l} < \hat{\beta}_{i_k}, l = 1, 2, \ldots, k-1] \\
&= \Pr[Z_{i_l} < D_{i_l}, l = 1, 2, \ldots, k-1] \\
&\geq \Pr[Z_{i_l} < Q_{i_l}, l = 1, 2, \ldots, k-1].
\end{aligned}$$

The inequality follows because n_{i_l} (the sample size of system i_l) is determined such that $n_{i_l} \geq \max((h_L S_{i_l,0}/d_a^*)^2, (h_L S_{i_l,0}^2/d_a^*)^2/2)$ and $D_{i_l} \geq Q_{i_l}$; see [John and Chen (2006)]. Note that D_{i_l} and Q_{i_l} can be considered as the one-tailed $(P^*)^{1/(k-1)}$ confidence interval half width of Z_{i_l}.

Let Φ denote the cdf of the standard normal distribution. Then

$$\begin{aligned}
&\Pr[Z_{i_l} < Q_{i_l}, l = 1, 2, \ldots, k-1] \\
&= E[\Pr[Z_{i_l} < Q_{i_l}, l = 1, 2, \ldots, k-1 | S_{i_1,0}^2, S_{i_2,0}^2, \ldots, S_{i_k,0}^2]] \\
&\geq E[\Pi_{l=1}^{k-1} \Phi(Q_{i_l})].
\end{aligned}$$

The inequality follows from Slepian's inequality [Tong (1980)] because Z_{i_l} for $l = 1, 2, \ldots, k-1$ are correlated. Furthermore, under the LFC and conditioning on $S_{i_1,0}^2, S_{i_2,0}^2, \ldots, S_{i_k,0}^2$, Z_{i_l} follows $N(0,1)$ asymptotically.

Let ω denote the pdf of the χ^2 distribution with $n_0 - 1$ d.f. Let

$$Y = \frac{h_L}{\sqrt{(n_0 - 1)(1/x + 1/y) + (n_0 - 1)^2(1/x^2 + 1/y^2)}}.$$

Based on the property that the variables $\chi_{i_l} = (n_0 - 1)S_{i_l,0}^2/\sigma_{i_l}^2$, $l = 1, 2, \ldots, k$ are independent χ^2 variables with $n_0 - 1$ d.f., it can be shown that

$$E[\Pi_{l=1}^{k-1} \Phi(Q_{i_l})] =$$

$$E\left[\Pi_{l=1}^{k-1} \Phi\left(\frac{h_L}{\sqrt{(n_0 - 1)(1/\chi_{i_l} + 1/\chi_{i_k}) + (n_0 - 1)^2(1/\chi_{i_l}^2 + 1/\chi_{i_k}^2)}}\right)\right]$$

and

$$\mathrm{P(CS)} \geq \int_0^\infty \left[\int_0^\infty \Phi(Y)\,\omega(x)dx\right]^{k-1} \omega(y)dy.$$

We set the right-hand side to P^* and solve for h_L. Unfortunately, analytical solutions to compute h_L from the above equation are difficult to obtain. We use another property of h_L to estimate its values. It can be shown that

$$\begin{aligned}
&\mathrm{P(CS)}\\
&\geq \Pr[Z_{i_l} < Q_{i_l}, l = 1, 2, \ldots, k-1]\\
&= \Pr[Z_{i_l}\sqrt{\sigma_{i_l}^2/S_{i_l,0}^2 + \sigma_{i_l}^4/S_{i_l,0}^4 + \sigma_{i_k}^2/S_{i_k,0}^2 + \sigma_{i_k}^4/S_{i_k,0}^4} < h_L,\\
&\quad l = 1, 2, \ldots, k-1]\\
&= \Pr[Z_{i_l}\sqrt{(n_0-1)(1/\chi_{i_l} + 1/\chi_{i_k}) + (n_0-1)^2(1/\chi_{i_l}^2 + 1/\chi_{i_k}^2)} < h_L,\\
&\quad l = 1, 2, \ldots, k-1].
\end{aligned}$$

Let

$$\Upsilon = \max_{l=1}^{k-1} Z_{i_l}\sqrt{(n_0-1)(1/\chi_{i_l} + 1/\chi_{i_k}) + (n_0-1)^2(1/\chi_{i_l}^2 + 1/\chi_{i_k}^2)},$$

then

$$\mathrm{P(CS)} \geq \Pr[\Upsilon < h_L] = P^*.$$

That is, h_L is the P^* quantile of the distribution of the variable Υ under the LFC. Consequently, we can use simulation procedures of estimating quantile to estimate the value of h_L. This approach is easier than the numerical integration because quantile estimates of independent and identically distributed (i.i.d.) samples can be easily obtained through order statistics without any complicated operations.

Note that variables Z_{i_l} for $l = 1, 2, \ldots, k-1$ are correlated, and the correlation coefficients are unknown. Let ζ_{i_l} for $l = 1, 2, \ldots, k-1$ be i.i.d. N(0,1) variables and

$$\tau = \max_{l=1}^{k-1} \zeta_{i_l}\sqrt{(n_0-1)(1/\chi_{i_l} + 1/\chi_{i_k}) + (n_0-1)^2(1/\chi_{i_l}^2 + 1/\chi_{i_k}^2)}.$$

Follow from Slepian's [Tong (1980)] inequality

$$\mathrm{P(CS)} \geq \Pr[\Upsilon < h_L] \geq \Pr[\tau < h_L].$$

A cost-effective sequential approach to select the best system from k lognormal populations.

The Sequential Lognormal Selection algorithm:

(1) Initialize the set I to include all k lognormal populations. Simulate n_0 samples for each system $i \in I$. Set the iteration number $r = 0$, and $N_{1,r} = N_{2,r} = \ldots = N_{k,r} = n_0$, where $N_{i,r}$ is the sample size allocated for system i in the rth iteration.
(2) Compute the incremental number of samples for system i in the $(r+1)^{th}$ iteration $\delta_{i,r+1} = \lceil(\max((h_L S_{i,r}/\hat{d}_{i,r})^2, (h_L S_{i,r}^2/\hat{d}_{i,r})^2/2) - N_{i,r})^+/2\rceil$, where $\hat{d}_{i,r} = \max(d_a^*, L(\hat{\beta}_{b,r}) - \hat{\beta}_{i,r})$.
(3) If $\delta_{i,r+1} = 0$ and $i \neq b$ (where $\hat{\beta}_{b,r} = \max_{i \in I} \hat{\beta}_{i,r}$), remove system i from the subset I.
(4) If there is only one element in the subset I, go to step 6.
(5) Simulate $\delta_{i,r+1}$ additional samples for each system $i \in I$ in the $(r+1)^{th}$ iteration. Set $r = r + 1$ and go to step 2.
(6) Return the values b and $\hat{\beta}_b$, where $\hat{\beta}_b = \max_{i=1}^{k} \hat{\beta}_i$.

10.4 Other Approach of Selection Procedures

The indifference-zone selection procedures described so far attempt to minimize the sample sizes given the required minimal P(CS). Another goal is to maximize the P(CS) given a fixed sample size, e.g., OCBA (Optimal Computing Budget Allocation, see Section 11.2.3 and [Chen et al. (2000)]).

[Chen (2004); Chen and Kelton (2005)] compare the approach of OCBA and the enhanced indifference-zone selection procedures (when the underlying variables are normally distributed). The resulting ratios of the allocated sample sizes of OCBA and the enhanced selection procedures are close (when the underlying variables are normally distributed). In fact, the problem of optimizing P(CS) is the dual of optimizing sample sizes (without the indifference-zone approach) and has similar solutions.

Another approach to solve ranking and selection is to use the property of Brownian motion process (see, e.g., [Andradóttir and Kim (2010)]). They call this kind of procedures "fully sequential procedure." The procedures increase one sample of each system that is still under consideration at each iteration and eliminate systems from further simulation as procedures proceed. These procedures are efficient in terms of sample sizes, but not necessarily in terms of runtime.

Furthermore, several approaches have been proposed to select systems when there are multiple objectives, which will be discussed in Chapter 14.

Table 10.1 $\hat{P}$(CS) and sample sizes for experiment 1

	$P^* = 0.90$			$P^* = 0.95$		
Procedure	$\hat{P}$(CS)	$\overline{T}$	std(T)	$\hat{P}$(CS)	$\overline{T}$	std(T)
SRS(20)	0.9925	1243	268	0.9956	1567	309
SRS(30)	0.9946	1221	253	0.9969	1517	294

10.5 Empirical Experiments

In this section, we present some empirical results of using the proposed framework to select the best system(s) from normal, exponential, and log-normal populations.

10.5.1 *Experiment 1: Normal Populations*

There are ten alternative designs in the selection subset. Suppose $X_{ij} \sim N(i, 6^2)$, $i = 1, 2, \ldots, 10$, where $N(\mu, \sigma^2)$ denotes the normal distribution with mean μ and variance σ^2. We want to select a design with the minimum mean: design 1. The indifference amount d^* is set to 0.90 for all cases. We list the actual P(CS) of the SRS procedure. We use two different initial number of replications, $n_0 = 20$ and 30. Furthermore, 10,000 independent experiments are performed to estimate the actual P(CS) by $\hat{P}$(CS): the proportion of the 10,000 experiments in which we obtained the correct selection.

The results of experiment 1 are summarized in Table 10.1. The $\hat{P}$(CS) column lists the proportion of correct selection. The $\overline{T}$ column lists the average of the total number of simulation replications ($\overline{T} = \sum_{r=1}^{10000} \sum_{i=1}^{10} T_{r,i}/10000$, and $T_{r,i}$ is the number of total replications or batches for design i at the rth simulation run) used in each procedure. The std(T) column lists the standard deviation of the number of total sample sizes T at each independent simulation run. The SRS(20) row lists the results of the procedure executed with initial replications $n_0 = 20$ (and similarly for $n_0 = 30$).

The $\hat{P}$(CS)'s are all larger than the specified P^*. Because the variance of the sample is larger with a smaller initial sample size n_0, the SRS procedure allocates more samples with smaller n_0.

10.5.2 *Experiment 2: Exponential Populations*

Let expo(β) denote the exponential distribution with parameter β. There are $k = 10$ alternatives with distribution expo(β_i) for $i = 1, 2, \ldots, k$ and the indifference amount is $d_r^* = 1.4$. Note that the parameter β of the exponential distribution is a scale parameter. Hence, the relative indifference zone should be used. The objective is to select a subset of size 3 containing the best 3 alternatives.

To estimate the h value, set $\beta_{i_l} = 1.0$ for $l = 1, 2, 3$ and $\beta_{i_l} = 1.4$ for $l = 4, 5 \ldots, k$. Let $\bar{X}_{i_l}(n)$ for $l = 1, 2, \ldots, k$ be the sample means obtained with n samples from each system. Let $\bar{X}_c = \max_{l=1}^{3} \bar{X}_{i_l}(n)$ and $\bar{X}_u = \min_{l=4}^{10} \bar{X}_{i_l}(n)$. When $n = 1$, the 0.01786 quantile of the distribution of $\varphi = \bar{X}_c/\bar{X}_u$ is 1.0. In other words, with only one observation from each system the P(CS) is only 0.01786. From these experiments, it is observed that the 0.90 quantile of the random variable φ (i.e., the h value) is approximately 0.999378 when $n = 108$. That is, $\Pr[\bar{X}_c < 0.999378\bar{X}_u] \approx 0.90$. Consequently, $\Pr[\bar{X}_c < \bar{X}_u]$ will be slightly greater than 0.90.

The subset selection procedure is then tested under the LFC, i.e., assuming $\beta_{i_l} = 1.0$ for $l = 1, 2, 3$ and $\beta_{i_l} = 1.4$ for $l = 4, 5, \ldots, k$. In each replication, 108 samples are simulated to compute the sample means for each system. The three systems having the smallest sample means are then selected as the best three systems. 10,000 independent replications are performed and the observed P(CS) is the proportion of independent replications that the procedure correctly selected systems i_l for $l = 1, 2, 3$. The observed P(CS) is 0.8990, slightly below the nominal value of 0.90.

The selection was also performed with the two-stage procedure described in Section 10.2.3, i.e., the ratios of the allocated sample sizes are similar to those of Eq. (10.5). The total sample size was fixed to 1080 with the initial sample size $n_0 = 20$. The observed P(CS) is 0.9145, slighter higher than the single-stage procedure.

10.5.3 *Experiment 3: Lognormal Populations*

In this experiment, we test the procedure with the systems under the LFC, i.e., $\beta_1 = \beta_2 = \beta_3 - d_a^* = 1.0$. We chose the first-stage sample size to be $n_0 = 5, 15, 25, 35$ and 45. The number of systems under consideration is $k = 3$. The indifference amount, d_a^*, is set to 0.5. The required minimal P(CS), P^*, is set to 0.90. Table 10.2 lists the configuration of the mean and variance. We test the LFC because the minimal P(CS) occurs under

Table 10.2 Parameter configuration of experiment 3

Parameter	μ_1	μ_2	μ_3	σ_1	σ_2	σ_3
Value	-0.125	-0.28	0.055	1.5	1.6	1.7

Table 10.3 The observed P(CS) with $P^* = 0.90$ and $k = 3$

	JC				SLS				
n_0	$\overline{T}_1$	$\overline{T}_2$	$\overline{T}_3$	$\hat{P}$(CS)	$\overline{T}_1$	$\overline{T}_2$	$\overline{T}_3$	$\hat{P}$(CS)	Iter
5	620	735	907	0.949	394	529	752	0.985	10
15	173	213	276	0.930	117	152	250	0.961	9
25	146	180	229	0.929	104	134	210	0.933	8
35	134	167	213	0.935	101	123	196	0.933	8
45	126	162	205	0.945	99	120	190	0.920	8

this configuration. Note that the greatest benefit of the SLS (Sequential Lognormal Selection) procedure will be fully realized under the non-LFC.

Table 10.3 lists the results from [John and Chen (2006)] (denoted JC) procedure and the SLS procedure. The n_0 column is the initial sample size. The $\overline{T}_i$ columns list the average of the sample sizes allocated for system i. The $\hat{P}$(CS) column lists the observed P(CS), i.e., the proportion that system 3 is selected. The iter column lists the average number of iterations of the SLS procedure. These results are based on 1,000 independent simulation runs. All observed P(CS)'s are greater than the specified nominal value of 0.90, which indicates that these procedures are conservative even under the LFC. The SLS procedure generally achieves the specified P(CS) with smaller sample sizes than the JC procedure. Furthermore, the invoked number of iterations by the SLS procedure is small.

It seems that the SLS procedure is likely to over-estimate the variance with a smaller initial sample size, thereby allocating larger than necessary sample sizes and achieving the P(CS) that is much greater than the nominal value. As the initial sample size increases, the variance estimates become more accurate and the observed P(CS)'s become closer to the nominal value. Moreover, as variance estimates become more accurate, the difference of the allocated sample sizes between SLS and JC becomes smaller.

Under the LFC, the reduction of sample sizes is mainly the result of sequentializing the procedure because $\hat{d}_{lk} \approx d^*_a$. The sequentialized procedure reduces the frequency of over-allocating samples.

10.6 Summary

This chapter has presented a framework for selecting the best system or a subset of the best systems. The framework can be applied to the normal and lognormal populations and potentially provide helpful insights and additional capabilities to existing ranking and selection processes. In cases that the underlying distributions are unknown, users can use batch means to "manufacture" approximately i.i.d. normal samples and apply selection procedures that are developed for normal populations. The efficiency of selection procedures can be increased by using variance reduction techniques, e.g., common random numbers.

Chapter 11

Computing Budget Allocation of Selection Procedures

One crucial element in R&S procedures is the event of "correct selection" (CS) of the true best system. In a stochastic simulation, the possibility of CS, denoted by P(CS), increases as the sample sizes become larger. Most *indifference-zone-selection* procedures are directly or indirectly based on the procedures of [Dudewicz and Dalal (1975)] and [Rinott (1978)]. However, these procedures determine the number of additional replications based on a conservative LFC assumption and do not take into account the value of sample means. If the accuracy requirement is high and the total number of designs in a decision problem is large, then the total simulation cost can easily become prohibitively high.

Some new approaches, such as *Optimal Computing Budget Allocation* (OCBA) [Chen et al. (2000)] and the *Enhanced Two-Stage Selection* (ETSS) procedure [Chen and Kelton (2005)], incorporate first-stage sample mean information with sample variance in determining the number of additional replications. In numerical testing, both OCBA and ETSS demonstrate a significant reduction in computing effort compared to Rinott's procedure. The basic idea of those procedures is that to ensure a high probability of correctly selecting a good design, a larger portion of the computing budget should be allocated to those designs that are critical in the process of identifying good designs. Overall simulation efficiency is improved as less computational effort is spent on simulating non-critical designs and more is spent on critical designs.

In this chapter, we investigate the sample size allocation strategy of several selection procedures. We focus on those procedures which intend to allocate simulation trials to designs in a way that maximizes P(CS) within a given computing budget. Other researches have previously examined various approaches for efficiently allocating a fixed computing budget across design alternatives, in particular [Chen et al. (2000)]. Traditional indifference-zone-selection procedures and optimal computing budget allocations have been treated as two completely separate approaches. [Chen and Kelton (2005)] demonstrate that these two classes of approaches obtain very similar results. Based on those findings, we examine the relationship between P(CS) and sample size allocation strategy and develop an approach for solving the budget allocation problem. The new procedure not only offers analyst more options without losing anything in terms of computational/statistical efficiency but also provides some intuition and insight of existing procedures. For sequential procedures to work efficiently, a good incremental sample size must be used. We develop a new strategy to calculate the incremental size dynamically at each iteration to further improve the efficiency of these procedures. The proposed approach is simple, general, practical and complementary to other techniques.

11.1 Problem Statement

There exists a large literature on assessing P(CS) based on classical statistical models. To facilitate the derivation of our approximation of P(CS), we assume the means and variances are known. Let $\phi(x)$ and $\Phi(x)$ denote the probability density and distribution function, respectively, of the standard normal distribution. Let $\delta_{i_l} = \mu_{i_l} - \mu_{i_1}$ for $l = 2, 3, \ldots, k$, and $\delta_{i_1} = \delta_{i_2}$. Then

$$\begin{aligned} \mathrm{P(CS)} &= \mathrm{P}[\bar{X}_{i_1} < \bar{X}_{i_l}, \text{ for } l = 2, 3, \ldots, k] \\ &= \mathrm{P}[\bar{X}_{i_1} - \bar{X}_{i_l} + \delta_{i_l} < \delta_{i_l}, \text{ for } l = 2, 3, \ldots, k] \\ &\geq \Pi_{l=2}^{k} \mathrm{P}[\bar{X}_{i_1} - \bar{X}_{i_l} + \delta_{i_l} < \delta_{i_l}] \\ &= \Pi_{l=2}^{k} \Phi(\delta_{i_l}/\sqrt{\sigma_{i_l}^2/N_{i_l} + \sigma_{i_1}^2/N_{i_1}}) \\ &= \Pi_{l=2}^{k} \Phi(Y_{i_l}). \end{aligned}$$

The inequality follows from Slepian's inequality [Tong (1980)] since the values $\bar{X}_{i_1} - \bar{X}_{i_l}$ are positively correlated. The second to last equality follows from the fact that the variate

$$Z_{i_l} = \frac{\bar{X}_{i_1} - \bar{X}_{i_l} + \delta_{i_l}}{\sqrt{\sigma_{i_l}^2/N_{i_l} + \sigma_{i_1}^2/N_{i_1}}}$$

has a $N(0,1)$ distribution. The last equality follows since

$$Y_{i_l} = \delta_{i_l}/\sqrt{\sigma_{i_l}^2/N_{i_l} + \sigma_{i_1}^2/N_{i_1}}.$$

[Chen (2004)] develops a normal approximated two-stage selection procedure (NTSS). The result is summarized as follows:

Theorem 11.1. *For k competing designs whose performance measure X_{ij} are normally distributed with means $\mu_1, \mu_2, \ldots, \mu_k$ and unknown variances that need to be estimated by sample variances $S_1^2(r), S_2^2(r), \ldots, S_k^2(r)$, where r is the current sample size, $P(CS)$ will be at least P^* when the sample size for design i is*

$$N_i = \max(r, \lceil (h_t S_i(r)/d_i)^2 \rceil), \text{ for } i = 1, 2, \ldots, k,$$

where the critical value $h_t = \sqrt{2} t_{P,r-1}$, $P = (P^*)^{1/(k-1)}$, and $d_i = \max(d^*, \mu_i - \mu_{i_1})$.

When $\mu_{i_2} \geq \mu_{i_1} + d^*$, the allocated sample sizes guarantee $\Phi(Y_{i_l}) \geq (P^*)^{1/(k-1)}$ for $l = 2, 3, \cdots, k$. Hence $\text{P(CS)} \geq \Pi_{l=2}^{k} \Phi(Y_{i_l}) \geq P^*$. In practice, however, the true means are unknown, sample means are used to estimate d_i, e.g., Eq. (10.1).

11.2 A Heuristic Computing Budget Allocation Rule

While the allocated sample sizes in the previous section guarantee P(CS) $\geq P^*$ when true means are known, we don't know whether they are optimal. We aim to derive sample size allocation rules by considering the following optimization problem:

$$\min \sum_{i=1}^{k} N_i$$
$$\text{subject to}$$
$$\prod_{l=2}^{k} \Phi(Y_{i_l}) \geq P^*$$
$$N_i \in N, i = 1, 2, \ldots, k.$$

Here N is the set of non-negative integers and $\sum_{i=1}^{k} N_i$ denotes the total computational cost assuming the simulation times for different designs are roughly the same. That is, we want to minimize the sample sizes that achieve the specified minimal P(CS), P^*. Because of the complexity there is no known analytical solutions of this optimization problem. It will be better if we can obtain analytical solution of the above optimization problem. However, $\prod_{l=2}^{k} \Phi(Y_{i_l})$ is a lower bound of P(CS). It is not clear whether the optimal sample sizes subject to $\prod_{l=2}^{k} \Phi(Y_{i_l}) \geq P^*$ constraint is the optimal sample sizes subject to P(CS) $\geq P^*$ constraint. Moreover, even though we are interested in the solution of this deterministic problem, in practice this optimization is a stochastic programming problem since the means and the variances are unknown, therefore, are not deterministic, and need to be estimated by sample means and sample variances.

To obtain a near optimal solution we propose to *heuristically* decompose the above optimization problem into the following two formulations.

$$\text{I: min} \sum_{l=2}^{k} Y_{i_l}$$

$$\text{subject to}$$

$$\prod_{l=2}^{k} \Phi(Y_{i_l}) \geq P^*.$$

and

$$\text{II: min } N_{i_1} + N_{i_2}$$

$$\text{subject to}$$

$$Y_{i_2} \geq z_P$$

$$N_{i_1}, N_{i_2} \in N.$$

Recall that z_P is the P quantile of the standard normal distribution. Throughout the rest of the chapter, we assume the right hand side of the equations used to compute N_i results in integer; if it is not integer then the smallest integer that is greater than that value should be used. This decomposition is based on the formulation of assessing P(CS) in Section 11.1. We cannot even conjecture whether the optimizer of these two problems is the optimizer of the original problem, however, it is a reasonable approach to search for a near optimal solution. Furthermore, the purpose of budget allocation is to improve simulation efficiency, we need a relatively fast and inexpensive way of achieving P^* within the budget allocation procedure. Efficiency is more crucial than estimation accuracy in this setting.

We now show that I is optimized when $Y_{i_l} = z_P$, i.e., $\Phi(Y_{i_l}) = P$, for $l = 2, 3, \ldots, k$; where $P = (P^*)^{1/(k-1)}$. The Lagrangian relaxation function

$$L = \sum_{l=2}^{k} Y_{i_l} - \lambda\left(\prod_{l=2}^{k} \Phi(Y_{i_l}) - P^*\right).$$

The Karush-Kuhn-Tucker (KKT) conditions:

$$\frac{\partial L}{\partial Y_{i_l}} = 1 - \lambda \prod_{j=2, j\neq l}^{k} \Phi(Y_{i_j})\phi(Y_{i_l}) = 0, \text{ for } l = 2, 3 \ldots, k.$$

$$\frac{\partial L}{\partial \lambda} = \prod_{l=2}^{k} \Phi(Y_{i_l}) - P^* = 0.$$

The KKT conditions are satisfied when $Y_{i_l} = z_P$, for $l = 2, 3, \ldots, k$, hence, it must be the optimal solution.

The following shows that II is optimized when $N_{i_1} = (z_P/\delta_{i_2})^2(\sigma_{i_1} + \sigma_{i_2})\sigma_{i_1}$ and $N_{i_2} = (z_P/\delta_{i_2})^2(\sigma_{i_1} + \sigma_{i_2})\sigma_{i_2}$. The Lagrangian relaxation function

$$L = N_{i_1} + N_{i_2} - \lambda(\delta_{i_2}/\sqrt{\sigma_{i_1}^2/N_{i_1} + \sigma_{i_2}^2/N_{i_2}} - z_P).$$

The KKT conditions:

$$\frac{\partial L}{\partial N_{i_1}} = 1 - \lambda(\delta_{i_2}\sigma_{i_1}^2/(2N_{i_1}^2(\sqrt{\sigma_{i_1}^2/N_{i_1} + \sigma_{i_2}^2/N_{i_2}})^3)) = 0.$$

$$\frac{\partial L}{\partial N_{i_2}} = 1 - \lambda(\delta_{i_2}\sigma_{i_2}^2/(2N_{i_2}^2(\sqrt{\sigma_{i_1}^2/N_{i_1} + \sigma_{i_2}^2/N_{i_2}})^3)) = 0.$$

$$\frac{\partial L}{\partial \lambda} = \delta_{i_2}/\sqrt{\sigma_{i_1}^2/N_{i_1} + \sigma_{i_2}^2/N_{i_2}} - z_P = 0.$$

From the first two equations, we obtain

$$\frac{N_{i_1}}{N_{i_2}} = \frac{\sigma_{i_1}}{\sigma_{i_2}}.$$

Solve the third equation to obtain

$$N_{i_1} = (z_P/\delta_{i_2})^2(\sigma_{i_1} + \sigma_{i_2})\sigma_{i_1}, \tag{11.1}$$

and

$$N_{i_2} = (z_P/\delta_{i_2})^2(\sigma_{i_1} + \sigma_{i_2})\sigma_{i_2}. \tag{11.2}$$

Consequently, for $k > 2$ P(CS) $\geq \prod_{l=2}^{k} \Phi(Y_{i_l}) \geq P^*$ can be achieved when

$$N_{i_1} = \max_{l=2}^{k}(z_P/\delta_{i_l})^2(\sigma_{i_1} + \sigma_{i_l})\sigma_{i_1}$$

and

$$N_{i_l} = (z_P/\delta_{i_l})^2(\sigma_{i_1} + \sigma_{i_l})\sigma_{i_l}, \text{ for } l = 2, 3, \ldots, k. \tag{11.3}$$

To simplify the computation effort, we set

$$N_{i_1} = (z_P/\delta_{i_2})^2(\sigma_{i_1} + \sigma_m)\sigma_{i_1}, \tag{11.4}$$

where $\sigma_m^2 = \max_{i=1, i \neq i_1}^{k} \sigma_i^2$. Recall that $\delta_{i_1} = \delta_{i_2} \leq \delta_{i_l} = \mu_{i_l} - \mu_{i_1}$, for $l = 2, 3, \cdots, k$.

There are $k-1$ pairwise comparisons between the best design i_1 and the rest, therefore, the sample size N_{i_1} can be computed $k-1$ different ways. We compute N_{i_1} based on Eq. (11.4) to ensure it is no less than the maximum of those $k-1$ values. The allocated sample sizes achieve $\prod_{l=2}^{k} \Phi(Y_{i_l}) = P^*$ when $\sigma_i^2 = \sigma^2$ for $i = 1, 2, \cdots, k$ and $\delta_{i_l} = \delta$ for $l = 2, 3, \cdots, k$. For all other cases, the allocated sample sizes will result in $\prod_{l=2}^{k} \Phi(Y_{i_l}) > P^*$.

Theorem 11.2. *For k competing designs whose performance measure X_{ij} are normally distributed with means $\mu_1, \mu_2, \ldots, \mu_k$ and unknown variances that need to be estimated by sample variances $S_1^2(r), S_2^2(r), \ldots, S_k^2(r)$, where r is the current sample size, $P(CS)$ will be at least P^* when the sample size for design i_1 is*

$$N_{i_1} = \max(r, (t_{P,r-1}/d_{i_2})^2(S_{i_1}(r) + S_m(r))S_{i_1}(r)), \qquad (11.5)$$

where $S_m^2(r) = \max_{i=1, i\neq i_1}^{k} S_i^2(r)$, and for design i_l

$$N_{i_l} = \max(r, (t_{P,r-1}/d_{i_l})^2(S_{i_1}(r) + S_{i_l}(r))S_{i_l}(r)), \text{ for } l = 2, 3, \ldots, k.$$

where $P = (P^*)^{1/(k-1)}$, and $d_i = \max(d^*, \mu_i - \mu_{i_1})$.

We denote the procedure developed based on Theorem 11.2 Near Optimal Selection (NOS) procedure in the remainder of this chapter. Note that the sample sizes allocated by NOS are optimal when all k systems have the same variance and $d_i = d^*$, for $i = 1, 2, \cdots, k$. So far, d_i has been estimated by Eqs. (10.1) or (10.3). Since to achieve $\mathrm{P}[\bar{X}_{i_1} < \bar{X}_{i_2}] \geq P$, we only need to allocate sample sizes large enough so that the one-tailed P confidence interval half width $w_{i_1 i_2} \leq \mu_{i_2} - \mu_{i_1}$; see Section 10.1.1 and [Chen (2004)].

The allocated sample sizes are larger than necessary when $\delta_{i_2} = \mu_{i_2} - \mu_{i_1}$ is far in excess of d^*. We can eliminate this drawback by setting $d_{i_1} = d_{i_2} = \max(d^*, \mu_{i_2} - \mu_{i_1})$. Consequently, the allocated sample sizes are optimal when all k systems have the same variance and d_i, for $i = 1, 2, \cdots, k$, have the same value. When the true means are unknown, we set $\hat{d}_b = \min_{i=1, i\neq b}^{k} \hat{d}_i$ for design b (recall that $\bar{X}_b(r) = \min_{1\leq i\leq k} \bar{X}_i(r)$), i.e., $\hat{d}_i = \max(d^*, |\bar{X}_i(r) - \min_{j=1, j\neq i}^{k} \bar{X}_j(r)|)$.

Generally speaking, we can improve the efficiency of R&S procedures with a *pre-selection*. Subset pre-selection is a screening device that attempts to select a (random-size) subset of the k alternative designs that contains the best one. Inferior designs will be excluded from further consideration, reducing the overall simulation effort. In the NOS procedure, design i having $N_i = n_0$ is excluded from further consideration. Hence, like ETSS and NTSS, NOS also has an intrinsic subset pre-selection built-in and does not require performing a pre-selection separately; see [Chen (2004)].

11.2.1 *Confidence Interval Half-Width and Computing Budget*

[Chen and Kelton (2003)] discuss the relationship between c.i. half-width and selection procedures. The one-tailed P c.i. half width between designs i_1 and i_2

$$w_{i_1 i_2} = z_P \sqrt{\frac{\sigma_{i_1}^2}{N_{i_1}} + \frac{\sigma_{i_2}^2}{N_{i_2}}}.$$

Moreover, optimization problem II stays the same when the constraint $Y_{i_2} \geq z_p$ is replaced by $w_{i_1 i_2} \leq \mu_{i_2} - \mu_{i_1}$. We investigate the following optimization problem.

$$\begin{aligned}\text{III: } &\min w_{i_1 i_2}\\ &\text{subject to}\\ &N_{i_1} + N_{i_2} = T\\ &N_{i_1}, N_{i_2} \in N.\end{aligned}$$

Similarly, optimization problem III stays the same when the objective min $w_{i_1 i_2}$ is replaced by max Y_{i_2}, or max $\Phi(Y_{i_2})$. The following shows that $w_{i_1 i_2}$ is optimized when $N_{i_1} = \sigma_{i_1} T/(\sigma_{i_1} + \sigma_{i_2})$ and $N_{i_2} = \sigma_{i_2} T/(\sigma_{i_1} + \sigma_{i_2})$. The Lagrangian relaxation function

$$L = z_P \sqrt{\sigma_{i_1}^2/N_{i_1} + \sigma_{i_2}^2/N_{i_2}} - \lambda(N_{i_1} + N_{i_2} - T).$$

The KKT conditions:

$$\frac{\partial L}{\partial N_{i_1}} = -z_p \sigma_{i_1}^2/(2N_{i_1}^2 \sqrt{\sigma_{i_1}^2/N_{i_1} + \sigma_{i_2}^2/N_{i_2}}) - \lambda = 0.$$

$$\frac{\partial L}{\partial N_{i_2}} = -z_p \sigma_{i_2}^2/(2N_{i_2}^2 \sqrt{\sigma_{i_1}^2/N_{i_1} + \sigma_{i_2}^2/N_{i_2}}) - \lambda = 0.$$

$$\frac{\partial L}{\partial \lambda} = N_{i_1} + N_{i_2} - T = 0.$$

From the first two equations, we obtain

$$\frac{N_{i_1}}{N_{i_2}} = \frac{\sigma_{i_1}}{\sigma_{i_2}}.$$

Solve the third equation to obtain $N_{i_1} = \sigma_{i_1} T/(\sigma_{i_1} + \sigma_{i_2})$ and $N_{i_2} = \sigma_{i_2} T/(\sigma_{i_1} + \sigma_{i_2})$.

The solutions of the optimization problems II and III indicate that the ratio of the optimal sample sizes to obtain a specified probability of correct comparison is the same as that to obtain a minimized c.i. half-width with a given computing budget. Without loss of generality, assume $\sigma_{i_2}/\sigma_{i_1} = \gamma$ and $N_{i_1} + N_{i_2} = T$. The one-tailed P c.i. half width between designs i_1 and i_2 computed at the end of NOS is

$$w_O = z_p \sqrt{\frac{\sigma_{i_1}^2}{\frac{T}{1+\gamma}} + \frac{\sigma_{i_2}^2}{\frac{\gamma T}{1+\gamma}}} = z_p \sqrt{\frac{w_O^2}{(1+\gamma) z_P^2} + \frac{\gamma w_O^2}{(1+\gamma) z_P^2}},$$

computed at the end of NTSS is

$$w_N = z_p \sqrt{\frac{\sigma_{i_1}^2}{\frac{T}{1+\gamma^2}} + \frac{\sigma_{i_2}^2}{\frac{\gamma^2 T}{1+\gamma^2}}} = z_p \sqrt{\frac{w_N^2}{2 z_P^2} + \frac{w_N^2}{2 z_P^2}},$$

and results in $w_O \leq w_N$. Similarly, when $k = 2$ the sample size allocation strategy of NOS will result in obtaining greater value of Y_{i_2} and $\Phi(Y_{i_2})$, i.e., greater accuracy of pairwise comparison. This implies that the empirical distribution of $\bar{X}_{i_2} - \bar{X}_{i_1}$ has smaller variance when the sample sizes are allocated by NOS than when the sample sizes are allocated by NTSS.

NTSS allocates N_i so that $\sigma_i^2/N_i = \delta_i/(2z_P^2)$, for $i = 1, 2, \cdots, k$. Hence, for $l = 2, 3, \cdots, k$

$$w_{i_1 i_l} = z_P \sqrt{\frac{\delta_{i_2}^2}{2z_P^2} + \frac{\delta_{i_l}^2}{2z_P^2}} \leq z_P \sqrt{\frac{\delta_{i_l}^2}{2z_P^2} + \frac{\delta_{i_l}^2}{2z_P^2}} = \delta_{i_l}.$$

Let η_l be real numbers, it is possible to allocate N_{i_l} for $l = 3, 4, \cdots, k$ so that

$$w_{i_1 i_l} = z_P \sqrt{\frac{\delta_{i_2}^2}{2z_P^2} + \frac{(\delta_{i_l} + \eta_l)^2}{2z_P^2}} = \delta_{i_l}.$$

It can be shown that $\delta_{i_l} + \eta_l = \sqrt{2\delta_{i_l}^2 - \delta_{i_2}^2}$. Hence, if we set $d_{i_1} = d_{i_2} = \max(d^*, \mu_{i_2} - \mu_{i_1})$ and $d_{i_l} = \max(d^*, \sqrt{2\delta_{i_l}^2 - \delta_{i_2}^2}) = \max(d^*, \sqrt{2(\mu_{i_l} - \mu_{i_1})^2 - (\mu_{i_2} - \mu_{i_1})^2})$, for $l = 3, 4, \cdots, k$, then even though the allocated sample sizes may not be optimal, they obtains $\Phi(Y_{i_l}) = P$ and $\prod_{l=2}^{k} \Phi(Y_{i_l}) = P^*$.

Similarly, if we set N_{i_1} and N_{i_2} according to Eqs. (11.1) and (11.2), and for $l = 3, 4, \cdots, k$

$$N_{i_l} = (z_P \sigma_{i_l})^2 \frac{\sigma_{i_1} + \sigma_{i_2}}{(\sigma_{i_1} + \sigma_{i_2})\delta_{i_l}^2 - \sigma_{i_1}\delta_{i_2}^2},$$

in the NOS procedure, we obtain $\Phi(Y_{i_l}) = P$ and $\prod_{l=2}^{k} \Phi(Y_{i_l}) = P^*$.

11.2.2 *Maximizing Probability of Correction Selection with a Given Computing Budget*

Based on the finding in previous section, we treat the optimization problem of maximizing P(CS) as the dual of minimizing the sample sizes. That is, we *assume* the ratio of sample sizes that maximizing P(CS) with a given computing budget and that minimizing the sample sizes with a specified minimal P(CS) are the same. We derive the sample-sizes-allocation rules that maximize $\prod_{l=2}^{k} \Phi(Y_{i_l})$ by the computing the ratios of sample sizes allocated by Eqs. (11.3) and (11.4). Thus, given a total number of simulation samples T to be allocated to k competing designs and their known means and variances are $\mu_1, \mu_2, \ldots, \mu_k$, and $\sigma_1^2, \sigma_2^2, \ldots, \sigma_k^2$ respectively, $\prod_{l=2}^{k} \Phi(Y_{i_l})$ will be near optimal (in term of the allocated sample sizes) when

$$\frac{N_{i_1}}{N_{i_2}} = \frac{(\sigma_{i_1} + \sigma_m)\sigma_{i_1}}{(\sigma_{i_1} + \sigma_{i_2})\sigma_{i_2}}, \tag{11.6}$$

where $\sigma_m^2 = \max_{i=1, i \neq i_1}^{k} \sigma_i^2$, and

$$\frac{N_{i_l}}{N_{i_2}} = \left(\frac{\delta_{i_2}}{\delta_{i_l}}\right)^2 \frac{(\sigma_{i_1} + \sigma_{i_l})\sigma_{i_l}}{(\sigma_{i_1} + \sigma_{i_2})\sigma_{i_2}}, \; l \in \{2, 3, \ldots, k\}. \tag{11.7}$$

On the other hand, the ratio derived from NTSS is

$$\frac{N_{i_l}}{N_{i_2}} = \left(\frac{\delta_{i_2}}{\delta_{i_l}}\right)^2 \frac{\sigma_{i_l}^2}{\sigma_{i_2}^2}, \; l \in \{1, 2, \ldots, k\}. \tag{11.8}$$

If the variances are equal, i.e., $\sigma_i^2 = \sigma^2$ for $i = 1, 2, \cdots, k$, then Eqs. (11.6) and (11.7) can be simplified to Eq. (11.8), i.e., the allocated sample sizes are the same for NOS and NTSS.

When the objective is to maximize P(CS) with a given computing budget, we should not use the indifference amount d^* when allocating the sample sizes. If the indifference amount $d^* > \mu_{i_2} - \mu_{i_1}$ and the procedure evokes d^* in computing $\hat{d}_i$ (i.e. $\hat{d}_i = \max(d^*, |\bar{X}_i(r) - \min_{j=1, j \neq i}^{k} \bar{X}_j(r)|)$, then a larger portion of the computing budget will be allocated to inferior designs whose mean is far in excess of the best design and will result in suboptimal allocation.

We now present a cost-effective sequential approach based on the concept described earlier to select the best design from k alternatives with a given computing budget. In our procedure, we use mean and variances estimators $\bar{X}_i(r)$ and $S_i^2(r)$ to compute the ratios of Eqs. (11.6) and (11.7), and the estimator of δ_i is $\hat{d}_i = |\bar{X}_i(r) - \min_{j=1, j \neq i}^{k} \bar{X}_j(r)|$. We use the equation $S_i^2(r) = (\sum_j^r X_{ij}^2 / r - \bar{X}_i(r)^2) r/(r-1)$ to compute the variance estimator, therefore, we are only required to store the triple $(r, \sum_{j=1}^{r} X_{ij}, \sum_{j}^{r} X_{ij}^2)$, instead of the entire sequences $(X_{i1}, X_{i2}, \ldots, X_{ir})$.

Initially, n_0 simulation replications for each of the k designs are conducted to get some information about the performance of each design during the first stage. As simulation proceeds, the sample means and sample variances of each design are computed from the data already collected up to that stage. According to this collected simulation output, an incremental computing budget for each iteration, Δ_l is distributed to each design based on the ratios of Eqs. (11.6) and (11.7), where l is the iteration number. Ideally, each new replication should bring us closer to the optimal solution. The procedure will be iterated repeatedly until we have exhausted the pre-determined computing budget T. The algorithm is summarized as follows.

A Sequential Algorithm for Computing Budget Allocation:

(1) Simulate n_0 replications or batches for each design. Set $l = 0$, $N_1^l = N_2^l = \ldots = N_k^l = n_0$, and $T = T - kn_0$.
(2) Let $S_m^2(N_m^l) = \max_{i=1, i\neq b}^{k} S_i^2(N_i^l)$, $\bar{X}_b(N_b^l) = \min_{i=1}^{k} \bar{X}_i(N_i^l)$, $\bar{X}_s(N_s^l) = \min_{i=1, i\neq b}^{k} \bar{X}_i(N_i^l)$, $\hat{d}_i = |\bar{X}_i(N_i^l) - \min_{j=1, j\neq i}^{k} \bar{X}_j(N_j^l)|$. Set $l = l + 1$. Increase the computing budget (i.e., the number of additional simulations) by Δ_l and compute the new budget allocation, $N_1^l, N_2^l, \ldots, N_k^l$, such that

$$\frac{N_b^l}{N_s^l} = \frac{(S_b(N_b^{l-1}) + S_m(N_m^{l-1}))S_b(N_b^{l-1})}{(S_b(N_b^{l-1}) + S_s(N_s^{l-1}))S_s(N_s^{l-1})}, \tag{11.9}$$

and

$$\frac{N_i^l}{N_s^l} = \left(\frac{\hat{d}_s}{\hat{d}_i}\right)^2 \frac{(S_b(N_b^{l-1}) + S_i(N_i^{l-1}))S_i(N_i^{l-1})}{(S_b(N_b^{l-1}) + S_s(N_s^{l-1}))S_s(N_s^{l-1})},$$

$$i \neq b \text{ and } i \in \{1, 2, \ldots, k\}.$$

(3) Simulate additional $\max(0, N_i^l - N_i^{l-1})$ replications or batches for each design i, $i = 1, 2, \ldots, k$.
(4) $T = T - \Delta_l$. If $T > 0$, go to step 2.
(5) Return the values b and $\bar{X}_b(N_b^l)$, where $\bar{X}_b(N_b^l) = \min_{1\leq i\leq k} \bar{X}_i(N_i^l)$.

As simulation evolves, design b, which is the design with the smallest sample mean, may change from iteration to iteration, although it will converge to the optimal design as l goes to infinity. In addition, we need to select the initial number of simulations, n_0, and the increment, Δ_l, at each iteration. A suitable choice for n_0 is between 5 and 20 [Law (2014)]. Also, with a small Δ_l, we need to iterate the computation procedure in step 2 many times. On the other hand, with a large Δ_l, we are putting too much confidence on the mean and variance estimators of early iterations and can result in waste of computation time to obtain an unnecessarily high confidence level of non-critical designs. Instead of using a fixed Δ_l for every iteration, we suggest computing Δ_l dynamically at each iteration

$$\Delta_l = \min(T, \max(k, \lceil T/2 \rceil)). \tag{11.10}$$

Thus, the sequential procedure allocates incremental sample sizes aggressively at earlier iterations and become less aggressive as the procedure proceeds and the computing budget becomes scarce. This way we will be able to reduce the number of iterations of step 2 without the risk of putting too much resources to simulate non-critical designs.

In general, a smaller Δ_l will result in higher P(CS) with longer execution time. Hence, a conservative alternative to the above strategy is to increase sample size by one for each design having been allocated extra samples. For example, if the above sample-size-allocation rules have allocated extra $N_i^l - N_i^{l-1} (> 0)$ samples for design i, we set $N_i^l = N_i^{l-1} + 1$.

11.2.3 *Optimal Computing Budget Allocation (OCBA)*

[Chen et al. (2000)] propose OCBA that is based on a fixed total computing budget $T = \sum_{i=1}^{k} N_i$ and attempts to maximize *Approximate Probability of Correct Selection* (APCS). By the *Bonferroni* inequality [Law (2014)],

$$\begin{aligned}
\mathrm{P(CS)} &= \mathrm{P}[\bar{X}_{i_1} - \bar{X}_{i_l} + \delta_{i_l} < \delta_{i_l}, \text{ for } l = 2, 3, \ldots, k] \\
&\geq 1 - \sum_{l=2}^{k} (1 - \mathrm{P}[\bar{X}_{i_1} - \bar{X}_{i_l} + \delta_{i_l} < \delta_{i_l}]) \\
&= 1 - \sum_{l=2}^{k} (1 - \Phi(Y_{i_l})) \\
&= 2 - k + \sum_{l=2}^{k} \Phi(Y_{i_l}).
\end{aligned}$$

They consider the following optimization problem:

$$\max \text{APCS} : \sum_{l=2}^{k} \Phi\left(Y_{i_l}\right)$$
$$\text{subject to}$$
$$\sum_{i=1}^{k} N_i = T.$$
$$N_i \in N, i = 1, 2, \ldots, k.$$

They show that for a fixed number of replications or batches, the APCS can be asymptotically maximized when

$$\frac{N_i}{N_j} = \left(\frac{\sigma_i/\delta_{b,i}}{\sigma_j/\delta_{b,j}}\right)^2, i, j \in \{1, 2, \ldots, k\}, \text{ and } i \neq j \neq b,$$

$$N_b = \sigma_b \sqrt{\sum_{i=1, i\neq b}^{k} \frac{N_i^2}{\sigma_i^2}},$$

where $\delta_{b,i} = \bar{X}_i - \bar{X}_b$, $\bar{X}_b = \min_{1\leq i\leq k} \bar{X}_i$, and σ_i is the standard deviation of the response of design i. Since σ_i's are unknown, standard error $S_i(r)$'s are used to compute the sample sizes in the implementation. Note that when $k = 2$ the sample sizes allocated by OCBA satisfy the requirement $N_i/N_j = \sigma_i/\sigma_j$, i.e., the optimal solution of problem I.

While the NOS uses the maximum variance of the $k - 1$ system to compute the value of N_b, i.e., Eq. (11.4), the derivation of the results above assumes $N_b >> N_{i\neq b}$. However, it is not clear the results from OCBA imply the optimal sample sizes are reached when $\Phi\left(Y_{i_l}\right)$, for $l = 2, 3, \cdots, k$, are the same. Based on our experimental results, the allocated sample sizes from all three approaches are consistent with each other, i.e., the differences in performance are minor.

11.3 Empirical Experiments

[Chen et al. (2000)] present the numerical results of OCBA and other commonly used R&S procedures. They demonstrate that OCBA is more efficient in terms of sample size allocation. In this section we present some empirical results obtained from simulations using NTSS, NOS, and OCBA. Firstly, we compare the allocated sample sizes from NTSS and NOS to

Table 11.1 $\hat{P}$(CS) and sample sizes for experiment 1

	$P^* = 0.90$			$P^* = 0.95$		
Procedure	$\hat{P}$(CS)	$\overline{T}$	std(T)	$\hat{P}$(CS)	$\overline{T}$	std(T)
NTSS(20)	0.9520	1263	483	0.9641	1584	616
NOS(20)	0.9573	1313	475	0.9646	1638	606
NTSS(30)	0.9689	1272	413	0.9780	1550	524
NOS(30)	0.9718	1312	406	0.9786	1605	520

obtain the required minimal P(CS). Secondly, we compare the level of empirical P(CS) with given computing budgets. Furthermore, we use Eq. (11.10) to compute the incremental sample size at each iteration when we execute these procedures sequentially.

11.3.1 *Experiment 1 Equal Variances*

There are ten alternative designs in the selection subset. Suppose $X_{ij} \sim N(i, 6^2)$, $i = 1, 2, \ldots, 10$. We want to select a design with the minimum mean: design 1. We perform 10,000 independent experiments to estimate the actual P(CS) by $\hat{P}$(CS): the proportion of the 10,000 experiments in which we obtained the correct selection.

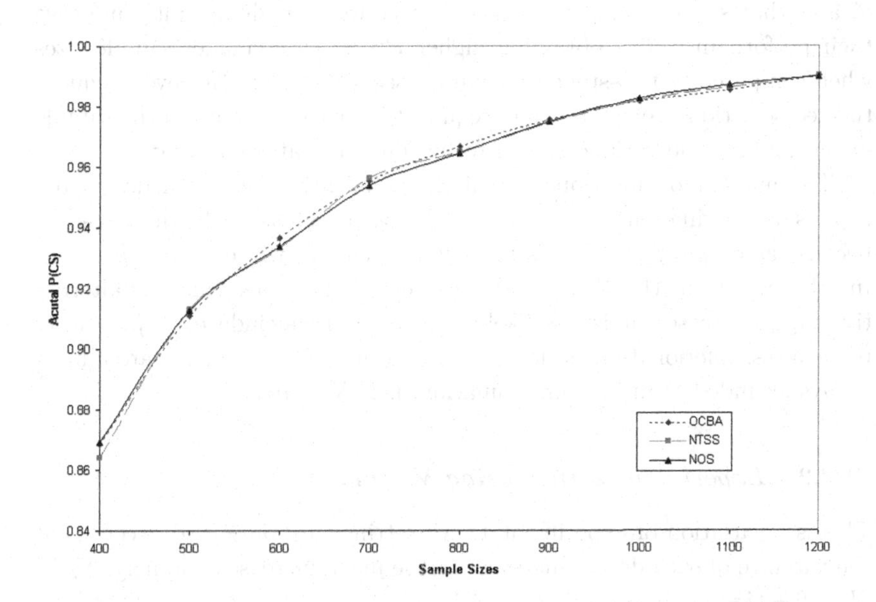

Fig. 11.1 $\hat{P}$(CS) and Sample Sizes for Experiment 1

The results of experiment 1 to minimize sample sizes are in Table 11.1. The $\hat{P}$(CS) column lists the proportion of correct selection. The $\overline{T}$ column lists the average of the number of total simulation replications ($\overline{T} = \sum_{R=1}^{10000} \sum_{i=1}^{10} T_{R,i}/10000$, and $T_{R,i}$ is the number of total replications or batches for design i at the Rth simulation run) used in each procedure. The std(T) column lists the standard deviation of the number of total simulation replications at each independent simulation run. The NTSS and NOS rows list the results of the respective procedures with initial replications $n_0 = 20$ (and similarly for $n_0 = 30$). All the $\hat{P}$(CS)'s are greater than the specified $P^* = 0.90$ and $P^* = 0.95$. Theoretically when the variances are equal among designs, NTSS and NOS are the same (see Section 11.2.2), however, in practice NOS uses the largest sample variance when computing N_b, i.e., Eq. (11.5). Hence, NOS generally obtains slightly higher P(CS) with slightly larger sample size than NTSS.

Figure 11.1 lists the results of experiment 1 when the objective is to maximize P(CS). We set the number of initial replications $n_0 = 20$. The computing budgets range from 400 to 1200 with increment size 100. The difference in $\hat{P}$(CS) is minor among these three procedures. The $\hat{P}$(CS) are greater than 0.98 when the given computing budget is 1200, which indicates that sequentializing selection procedures can significantly improve their performance, i.e., obtaining higher $\hat{P}$(CS) with smaller sample sizes when compared to two-stage procedures; see Table 11.1. However, sequentialized selection procedures may require longer run time since the sample means and sample variances need to be computed at each iteration.

The number of additional simulation replications for each design decreases as the differences $\hat{d}_{i \neq b} = \bar{X}_i - \bar{X}_b (> 0)$ increase. This makes sense because as $\hat{d}_{i \neq b}$ increases, it is more likely that we will conclude $\mu_i > \mu_b$. In other words, as the observed difference of sample means across alternatives $\hat{d}_{i \neq b}$ increases, it is less likely that we will conclude $\mu_i < \mu_b$. In all procedures, inferior designs, for instance designs 8 through 10, are almost always excluded from further simulation, i.e., $N_i \approx n_0$.

11.3.2 *Experiment 2 Increasing Variances*

This is a variation of experiment 1. All settings are preserved except that the variance of each design increases as the mean increases. Namely, $X_{ij} \sim N(i, (6 + (i - 1)/2)^2)$, $i = 1, 2, \ldots, 10$.

The results are in Table 11.2 and Figure 11.2. Since most designs have larger variances than in experiment 1, $\hat{P}$(CS) are not as good when the

Table 11.2 $\hat{P}$(CS) and sample sizes for experiment 2

	$P^* = 0.90$			$P^* = 0.95$		
Procedure	$\hat{P}$(CS)	$\overline{T}$	std(T)	$\hat{P}$(CS)	$\overline{T}$	std(T)
NTSS(20)	0.9477	1614	739	0.9526	2028	932
NOS(20)	0.9463	1793	713	0.9601	2279	916
NTSS(30)	0.9664	1534	607	0.9737	1920	774
NOS(30)	0.9672	1707	587	0.9761	2117	740

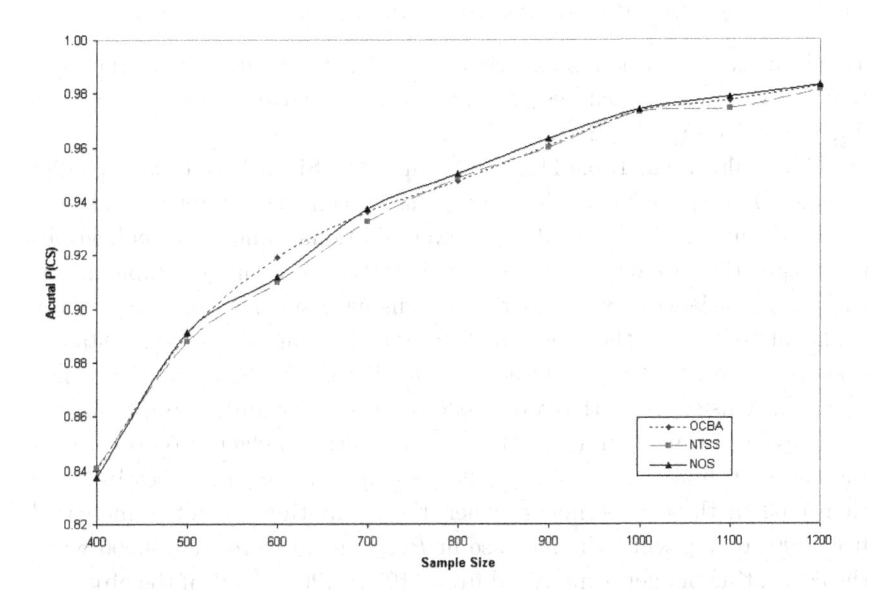

Fig. 11.2 $\hat{P}$(CS) and Sample Sizes for Experiment 2

computing budget is fixed. NOS has the best performance in this setting. All three procedures allocate relatively more additional simulation replications for designs with larger variances. These procedures take into consideration the difference between sample means, so $N_i < N_j$ even when $S_i^2(n_0) > S_j^2(n_0)$. [Chen and Kelton (2005)] indicate that procedures that take into account the difference between sample means have the most significant reduction in the number of replications or batches (compared to Rinott's procedure) when the inferior alternatives have larger variances. In such a case, Rinott's procedure tends to allocate most computing resource to inferior designs and so is inefficient.

Table 11.3 $\hat{P}$(CS) and sample sizes for experiment 3

	$P^* = 0.90$			$P^* = 0.95$		
Procedure	$\hat{P}$(CS)	$\overline{T}$	std(T)	$\hat{P}$(CS)	$\overline{T}$	std(T)
NTSS(20)	0.9583	1085	358	0.9685	1340	456
NOS(20)	0.9598	1089	363	0.9692	1346	459
NTSS(30)	0.9730	1112	303	0.9823	1350	384
NOS(30)	0.9694	1117	311	0.9788	1358	394

11.3.3 *Experiment 3 Decreasing Variances*

This is another variation of experiment 1. All settings are preserved except that the variance of each design decreases as the mean increases. Namely, $X_{ij} \sim N(i, (6 - (i-1)/2)^2)$, $i = 1, 2, \ldots, 10$.

The results are in Table 11.3 and Figure 11.3. Since designs have smaller variance, $\hat{P}$(CS) are better than in setting 1 when the computing budget is fixed. All three procedures allocate fewer additional simulation replications for designs that are clearly inferior in this setting, i.e., large sample means with small variances. Since inferior designs have smaller variances, we are confident to exclude those designs from further simulations. For instance, designs 7 through 10, are always excluded from further simulation, i.e., $N_i = n_0$. This suggests that we should use a smaller initial sample size.

These experiments indicate that the computing budget also experiences the effect of diminishing returns. For example, $\hat{P}$(CS) increases by more than 0.03 in these experiments when the computing budget is increased from 400 to 500, while the increase in $\hat{P}$(CS) is no more than 0.006 when the computing budget is increased from 1100 to 1200. Thus, if the objective is to minimize sample sizes and the given P^* is a small value, then analysts should consider a higher P^* since the marginal cost is small.

11.4 Summary

Traditional indifference-zone selection procedures are derived based on the LFC and are conservative. New approaches that take into account both the sample variances and the sample means can significantly improve the efficiency of selection procedures. We investigated the sample size allocation strategy of these procedures and developed a highly efficient procedure to identify a good design out of k alternatives. The purpose of this technique is to further enhance the efficiency of ranking and selection in simulation experiments. The objective is to maximize the simulation efficiency, expressed as P(CS) within a given computing budget. The incremental sample sizes

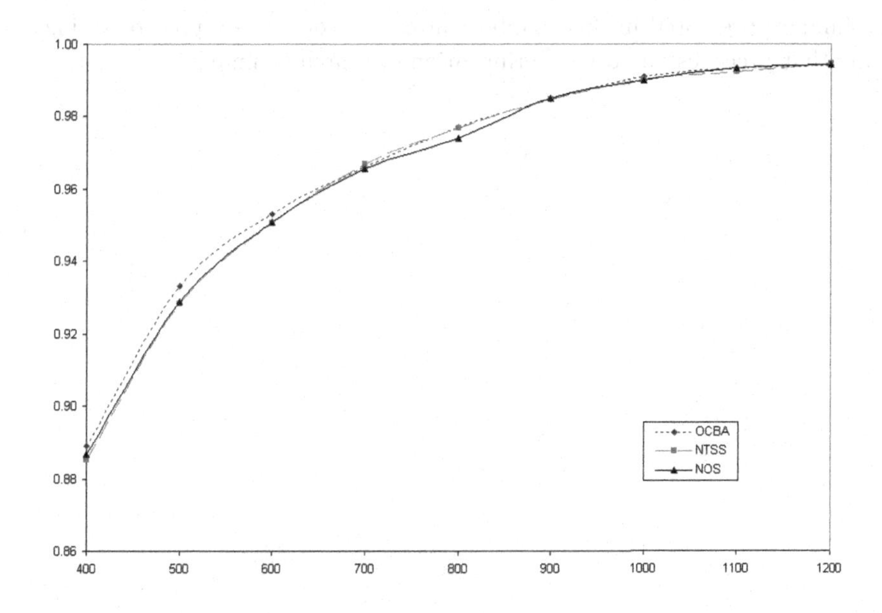

Fig. 11.3 $\hat{P}$(CS) and Sample Sizes for Experiment 3

at each iteration for each design are computed dynamically according to the sample means, the sample variances, and the available computing budget at each iteration. Our procedure allocates replications in such a way that optimally improves P(CS).

The performance differences among these three procedures are minor. Even though the incremental sample sizes for each design at each iteration are easier to compute in NTSS and NOS than OCBA, these formulas would be buried somewhere in the software so a little more simplicity of expression is not too important. However, the derivation of NTSS and NOS provides some insight of OCBA and computing budget allocation for selection. While it will result in sub optimal if NTSS and NOS evoke indifference amount in allocating sample sizes when the objective is to maximize P(CS), NTSS and NOS are able to explore the information of the indifference amount and estimate the required sample size for each design when the objective is to minimize computing budget given a required minimal P(CS), which can improve the computation efficiency. While ordinal optimization can converge exponentially fast, our simulation budget allocation procedure provides a way to further improve overall simulation efficiency. The techniques presented in this chapter can be considered as a pre-processing step that precedes any other optimization or search techniques.

Chapter 12

Using Common Random Numbers with Selection Procedures

The common random numbers (CRN) simulation technique is a variance-reduction method in which policy alternatives are tested against the same random input streams. The CRN literature suggests that positively correlated input streams will generate positively correlated policy responses and, therefore, that the variance of CRN estimators of response differences will be smaller than the variance of independent sample estimators. However, because the assumption of independent samples across systems is used to develop the two-stage selection procedures (see, e.g., [Dudewicz and Dalal (1975)]), CRN are generally not used with those procedures for variance reduction. There are other variance-reduction techniques, e.g., antithetic variates, control variates, see [Law (2014)] for details.

12.1 Common Random Numbers

In this section, we review the rational of using CRN. Consider the case of two alternative systems, where X_{1j} and X_{2j} are the observations from the first and second systems on the jth independent replication, and we want to estimate $\mu_1 - \mu_2$. If we make n replications of each system and let $Z_j = X_{1j} - X_{2j}$, for $j = 1, 2, \ldots, n$, then $E(Z_j) = \mu_1 - \mu_2$. It is known that $\bar{Z}(n) = \sum_{j=1}^{n} Z_j/n$ is an unbiased estimator of $\mu_1 - \mu_2$. Furthermore,

$$n\mathrm{Var}(\bar{Z}(n)) = \mathrm{Var}(Z_j) = \mathrm{Var}(X_{1j}) + \mathrm{Var}(X_{2j}) - 2\mathrm{Cov}(X_{1j}, X_{2j}).$$

If the simulations of the two different systems are carried out independently, i.e., with different random numbers, X_{1j} and X_{2j} will be independent, so that $\mathrm{Cov}(X_{1j}, X_{2j}) = 0$. On the other hand, if we carried out the simulations of systems 1 and 2 such that X_{1j} and X_{2j} are positively correlated, then $\mathrm{Cov}(X_{1j}, X_{2j}) > 0$, so that the variance of the estimator $\bar{Z}(n)$

is reduced. CRN is a technique used to induce this positive corvariance by using the same random numbers to simulation all systems. From an application perspective, using CRN allows us to compare different systems under similar circumstances. For an overview of CRN, please see [Law (2014)]).

In this chapter, we investigate the impact of using CRN in [Dudewicz and Dalal (1975)] procedure as well as its extension for subset selection. We show that it is generally safe to use CRN in these selection procedures even though they are derived based on the assumption of independent sampling. See [Nazzal et al. (2012)] for a practical example of applying CRN with indifference-zone selection procedures for simulation optimization.

12.2 The Basis of Correlated Order Statistics

Let E(X) and E(Y), respectively, denote the expected value of the random variables $X \sim f$ and $Y \sim g_{u:k}$. Here "$\sim$" denotes "is distributed as". The following are characteristics of the order statistics (of normal or t-distributed populations) that are verified with numerical analysis.

(1) If $u < (k+1)/2$, then $\mathrm{E}(Y) < \mathrm{E}(X)$.
(2) If $u = (k+1)/2$, then $\mathrm{E}(Y) = \mathrm{E}(X)$.
(3) If $u > (k+1)/2$, then $\mathrm{E}(Y) > \mathrm{E}(X)$.
(4) $\mathrm{Var}(Y) < \mathrm{Var}(X)$.
(5) The density of the mode of Y is larger than the density of the mode of X.
(6) If X's are correlated, $g_{u:k}$ converges to f as the corvariance become stronger.
(7) If X's are perfectly correlated (i.e., correlation coefficient is 1), $g_{u:k} = f$.

Because E(Y) changes as the covariance (of X) changes, we need to take into account the change in E(Y) in addition to the change in Var(Y) when considering the effect of using CRN in selection procedures.

12.2.1 *Using CRNs with Dudewicz and Dalal's Procedure*

Let $\mathrm{Pr}_C[E]$ and $\mathrm{Pr}_I[E]$, respectively, denote the probability of event E with and without CRN. Intuitively, $\mathrm{Pr}_I[T_{i_1} < T_{i_l} + h_1 \text{ for } l = 2,3,\ldots,k] < \mathrm{Pr}_C[T_{i_1} < T_{i_l} + h_1 \text{ for } l = 2,3,\ldots,k]$ because $\mathrm{Var}(T_{i_1} - T_{i_l})$ for $l = 2,3,\ldots,k$ is smaller with CRN.

We investigate the effect of using CRN in selection procedures from the order-statistics perspective. Let $T_u = \min_{l=2}^k T_{i_l}$. Then $T_u \sim g_{1:k-1}$, with f and F, respectively, are the pdf and cdf of the t distribution with $n_0 - 1$ d.f. Furthermore,

$$\mathrm{P(CS)} = \Pr[T_{i_1} - T_{i_l} \leq h_1 \text{ for } l = 2, 3, \ldots, k] = \Pr[T_{i_1} - T_u \leq h_1].$$

Recall that $\mathrm{Var}(T_{i_1} - T_u) = \mathrm{Var}(T_{i_1}) + \mathrm{Var}(T_u) - 2\mathrm{Cov}(T_{i_1}, T_u)$ and $\mathrm{Cov}(T_{i_1}, T_u) > 0$ when the CRN are properly synchronized. The standard deviation within each alternative from the first stage is not influenced by using CRN across systems. Thus, the required number of simulation replications or batches for each alternative should be consistent regardless of whether CRN are used. Let $\mathrm{Var}_I(T_u)$ and $\mathrm{Var}_C(T_u)$, respectively, denote the variance of T_u with independent sampling and with CRN. Depending on the strength of the correlation ($0 \leq \rho \leq 1$) among T_{i_l} for $l = 2, 3, \ldots, k$, $\mathrm{Var}_I(T_u) \leq \mathrm{Var}_C(T_u) \leq \mathrm{Var}(T_{i_1})$. Note that $\mathrm{Var}(T_u)$ is greater with CRN than without CRN and $\mathrm{Var}_C(T_u) = \mathrm{Var}(T_{i_1})$ when samples across systems are perfectly correlated, i.e., $\rho = 1$.

Because $\mathrm{Var}(T_u)$ is greater with CRN, $\mathrm{Var}(T_{i_1} - T_u)$ may be greater with CRN. Furthermore, if $\mathrm{Cov}(T_{i_1}, T_u) = 0$, then $\mathrm{Var}(T_{i_1} - T_u)$ will be greater with CRN. On the other hand, as discussed in Section 12.2, $\mathrm{E}(T_u) \leq \mathrm{E}(T_{i_1})$. Moreover, $\mathrm{E}(T_u) \to \mathrm{E}(T_{i_1})$ as the (positive) covariances become stronger, i.e., $\mathrm{E}(T_u)$ is increasing as the (positive) covariances become stronger. In the extreme, all $k-1$ observations are perfectly correlated, $\mathrm{E}(T_u) = \mathrm{E}(T_{i_1})$ because $g_{1:k-1} = f$. With this insight, it is clear that (as P^* deviates more from 1) even when $\mathrm{Cov}(T_{i_1}, T_u) = 0$ and $\mathrm{Var}(T_{i_1} - T_u)$ is greater with CRN, $\Pr_I[T_{i_1} < T_u] < \Pr_C[T_{i_1} < T_u]$. Recall that $\Pr[T_{i_1} > T_u]$ is roughly proportional to the area under the overlapping tail of the density functions T_{i_1} and T_u, see [Chen (2004)].

To show that CRN is valid in selection, we need to show that the P^* quantile of $T_{i_1} - T_u$ (i.e., the critical constant h_1) is smaller with CRN. That is, if $F_I(h_1) = F_C(h_c)$, then $h_c \leq h_1$, where $F_I(\cdot)$ and $F_C(\cdot)$, respectively, denote the cdf of the distribution of $T_{i_1} - T_u$ without and with CRN. Hence, $F_I(h_1) \leq F_C(h_1)$.

Figures 12.1 and 12.2 show the empirical density functions of $X_{i_1} - X_u$, where $X_{i_1} \sim N(0, 1)$ and X_u is the first-order statistics of 9 $N(0, 1)$ random variables. The graph on Figure 12.1 lists the pdf when X_{i_1} and X_u are correlated, while the graph on Figure 12.2 lists the pdf when X_{i_1} and X_u are independent. The covariances are 0, 0.5, and 0.95. Even though these graphs do not provide a rigorous proof, it does show that it is generally safe to use CRN in selection procedures to increase P(CS), especially when X_{i_1}

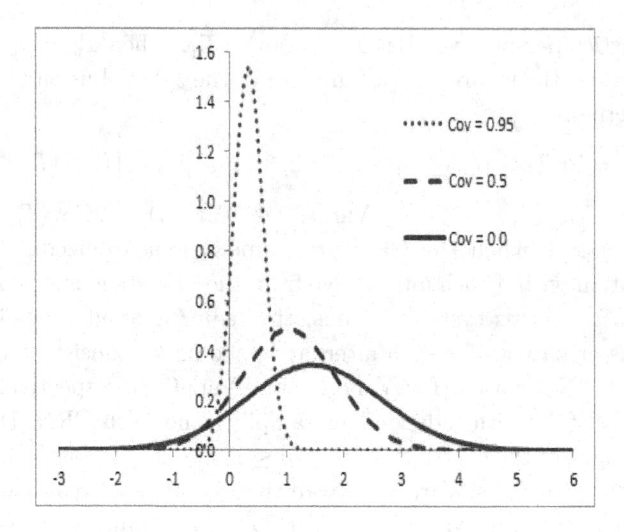

Fig. 12.1 Empirical density of $X_{i_1} - X_u$ of 10 $N(0,1)$ random variables

and X_u are correlated. In the cases that X_{i_1} and X_u are uncorrelated, it is not clear whether the pdf curves of correlated $X_{i_1} - X_u$ intersects the pdf curves of independent $X_{i_1} - X_u$ as $X \to \infty$. Nevertheless, it is clear that those pdf curves do not intersect when X is not in the extreme. Hence, as P^* decreases, the probability of using CRN "backfire" decreases.

12.2.2 *Subset Selection with CRN*

Let $\Pr_I[E]$ and $\Pr_C[E]$, respectively, denote the probability of event E without and with CRN. Recall that $T_c \sim g_{c:v}$, $T_u \sim g_{m-c+1:k-v}$, and $\Upsilon = T_c - T_u$. We intend to show that $\Pr_I[\Upsilon < h] \leq \Pr_C[\Upsilon < h]$. We begin our discussion with an example. If we are interested in the probability of correctly selecting a subset of size 5 containing 3 of the first 3 best from 10 alternatives, then $T_c \sim g_{3:3}(t_c)$ and $T_u \sim g_{3:7}(t_u)$. Furthermore, if the initial sample size is $n_0 = 20$, then f and F are, respectively, the pdf and cdf of the t-distribution with 19 d.f.

The distributions of T_c and T_u converge to T as the covariance becomes stronger. Consequently, the distribution of Υ converges to a single value 0 (i.e., variance is 0) as the covariance becomes stronger. Note that $\mathrm{E}(\Upsilon)$ may be non-positive. To investigate the effect of CRN to the distribution of $T_c - T_u$. We investigate the distribution of $X_c - X_u$, where $X_c \sim g_{c:v}$

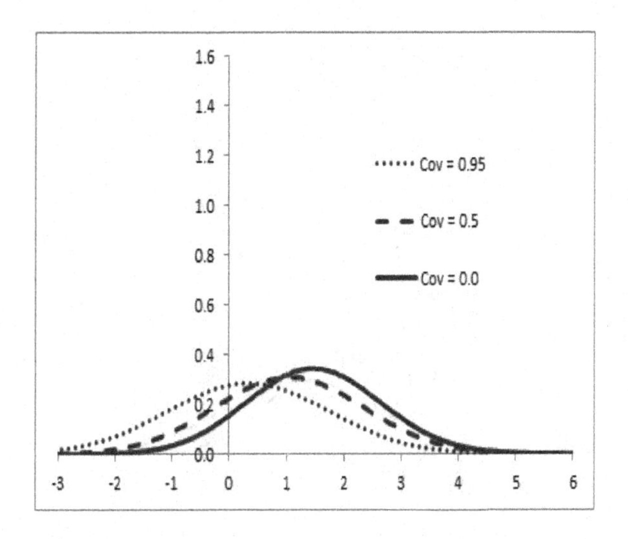

Fig. 12.2 Empirical density of $X_{i_1} - X_u$ of 10 $N(0, 1)$ random variables

and $X_u \sim g_{m-c+1:k-v}$ with f and F, respectively, being the pdf and cdf of the standard normal distribution. It is easier to manipulate the standard normal distribution than the t-distribution. Moreover, the pdf curves of $T_c - T_u$ and $X_c - X_u$ are similar.

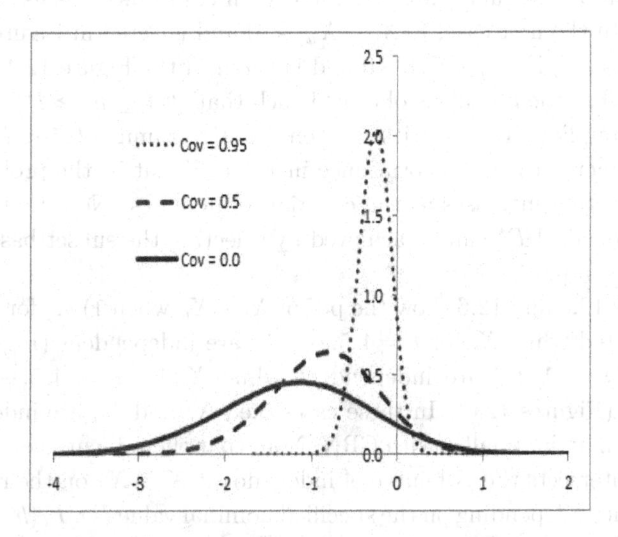

Fig. 12.3 Empirical probability densities of $X_c - X_u$

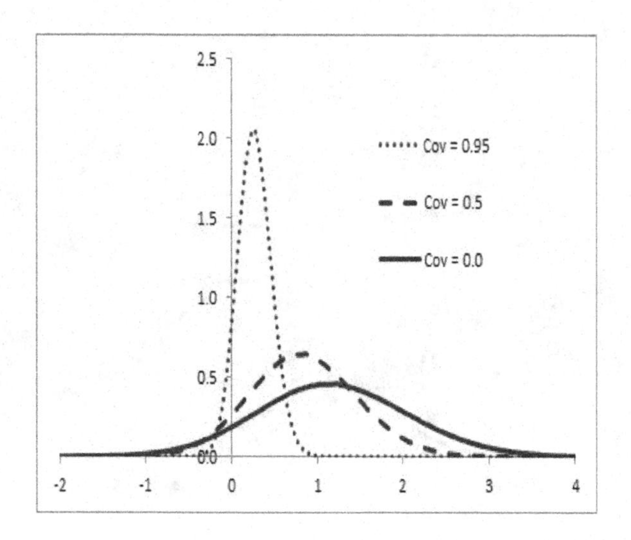

Fig. 12.4 Empirical probability densities of $X_c - X_u$

We investigate two cases: 1) $X_c \sim g_{1:3}$ and $X_u \sim g_{5:7}$ (i.e., $c = 1, v = 3, m = 5, k = 10$); 2) $X_c \sim g_{3:3}$ and $X_u \sim g_{3:7}$ (i.e., $c = 3, v = 3, m = 5, k = 10$). The empirical curves are generated with covariance equals to 0.0, 0.5, and 0.95. In the first case, $\mathrm{E}(X_c - X_u) < 0$ and is listed in Figure 12.3. In the second case, $\mathrm{E}(X_c - X_u) > 0$ and is listed in the Figure 12.4.

Recall that the h value is obtained such that $\mathrm{P}(\Upsilon \leq h) \geq P^*$ with independent sampling. Hence, with a given $h > 0$ (assuming $0.5 < P^* < 1.0$), $\mathrm{P}(\Upsilon \leq h)$ increases as the covariance increases. That is, the probability of correct selection increases as the covariance increases. Note that if $h \leq 0$, then the specified P^* can be achieved by selecting the subset based on the first-stage samples.

Figures 12.5 and 12.6 show the pdf of $X_c - X_u$ when 1) X_i for $i = 1, 2, 3$ are correlated while X_i for $i = 4, 5, \cdots, 10$ are independent (Figure 12.5); 2) X_i for $i = 1, 2, 3$ are independent while X_i for $i = 4, 5, \cdots, 10$ are correlated (Figures 12.6). In these cases, i.e., X_c and X_u are independent, the P(CS) may be smaller with CRN. Note that the pdf curve of correlated $X_c - X_u$ intersects the pdf curve of independent $X_c - X_u$ on the right tails. Consequently, depending on the specified nominal value P^*, $F_I(h) \geq F_C(h)$. Unfortunately, systems c and u are unknown. Consequently, to ensure that using CRN does not "backfire", all systems must be positively correlated.

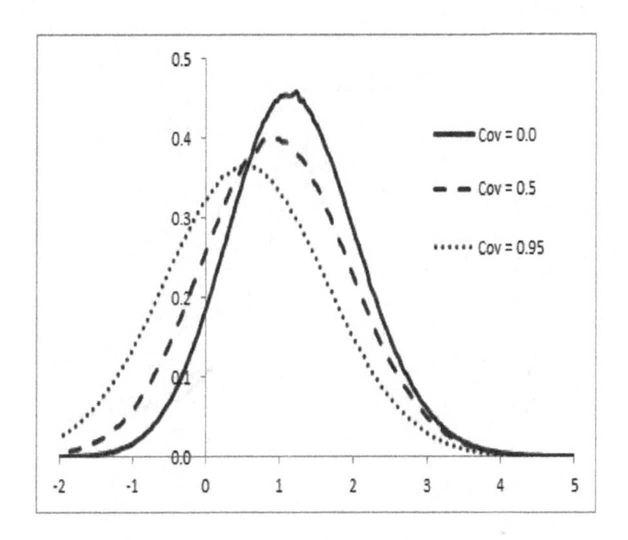

Fig. 12.5 Empirical probability densities of $X_c - X_u$

12.3 Empirical Experiments

In this section, we present some empirical results obtained from simulations using the P_E [Dudewicz and Dalal (1975)] and P_R [Rinott (1978)] procedures with CRN.

12.3.1 *Experiment 1: All Systems are Correlated*

We use the following techniques to create positively correlated random variates. Let $\xi_0 \sim N(0, \sigma_0^2)$, $\xi_i \sim N(\mu_i, 1 - \sigma_0^2)$ for $i = 1, 2, \ldots, k$ and let $Z_i = \xi_i - \xi_0$ for $i = 1, 2, \ldots, k$. Let $\varsigma_i \sim N(0, \sigma_0^2)$. It can be viewed as that $Z_i = \xi_i - \varsigma_i$ for $i = 1, 2, \ldots, k$ and ς_i is simulated with CRN. We set $\mu_1 = 0$ and $\mu_i = d^*$, for $i = 2, 3, \ldots, k$. Note that $Z_1 \sim N(0, 1)$ and $Z_i \sim N(d^*, 1)$ for $i = 2, 3, \ldots, k$ and correlates with

$$\mathrm{Cov}(Z_i, Z_j) = (\frac{1 - \sigma_0^2}{\sigma_0^2} + 1)^{-1/2}(\frac{1 - \sigma_0^2}{\sigma_0^2} + 1)^{-1/2} = \sigma_0^2,\ i \neq j.$$

We set $\sigma_0^2 = 0.00, 0.01, 0.05, 0.10, 0.25, 0.50, 0.75, 0.90$, $P^* = 0.95$, $n_0 = 20$, $k = 10$, and the indifferent amount $d^* = 0.01$. Note that $\xi_0 = 0$ when $\sigma_0^2 = 0.00$, hence, $Z_i = \xi_i$ for $i = 1, 2, \ldots, k$ are independent.

We consider the LFC, $\mu_1 + d^* = \mu_2 = \ldots = \mu_{10}$. The minimum P(CS) should occur at this configuration. Furthermore, 10,000 independent exper-

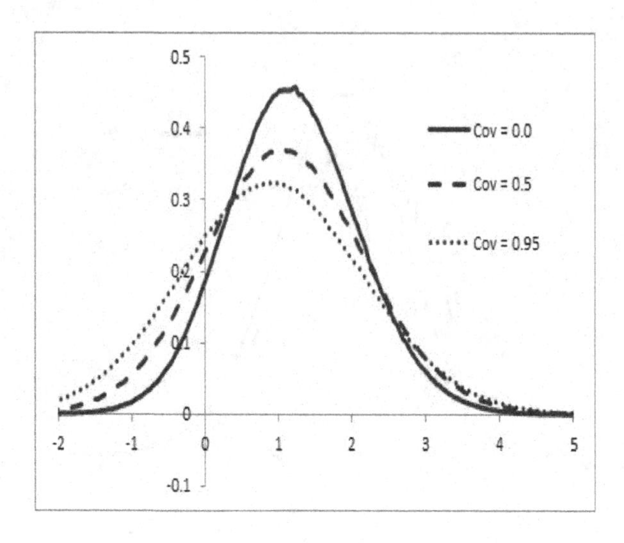

Fig. 12.6 Empirical probability densities of $X_c - X_u$

iments are performed to estimate the actual P(CS) by $\hat{P}$(CS), the proportion of the 10,000 experiments in which we obtained the correct selection.

Table 12.1 lists the results. We list the actual P(CS) of the P_E and P_R procedures with different values of σ_0^2. The $\overline{T}$ column lists the average of the total simulation replications ($\overline{T} = \sum_{r=1}^{10000} \sum_{i=1}^{k} N_{r,i}/10000$, $N_{r,i}$ is the number of total replications or batches for system i in the rth independent run) used in each procedure. As expected, the $\hat{P}$(CS)s increase as the covariance becomes stronger while the average sample sizes are about the same. The $\hat{P}$(CS)s of both P_E and P_R are slightly less than the nominal value when $\sigma_0^2 = 0.00$ and 0.01. We believe it is because of the stochastic nature of the experiments. Furthermore, the differences of the $\hat{P}$(CS)s of P_E and P_R are very small.

12.3.2 *Experiment 2: Best System is Independent with Others*

In this experiment, $Z_1 \sim N(0,1)$ and $Z_i \sim N(d^*,1)$ for $i = 2,3,\ldots,k$ and correlates with $\mathrm{Cov}(Z_i, Z_j) = \sigma_0^2$, $i \neq j, i \neq 1, j \neq 1$, while $\mathrm{Cov}(Z_1, Z_j) = 0$ for $j = 2,3,\ldots,k$. All other settings are the same as previous experiment.

Table 12.2 lists the results. As expected, the $\hat{P}$(CS)s are less than the corresponding $\hat{P}$(CS)s of experiment 1 because system 1 is independent of

Table 12.1 $\hat{P}$(CS) and sample sizes for experiment 1

Procedure/σ_0^2	0.00	0.01	0.05	0.10	0.25	0.50	0.75	0.90
P_E	0.9469	0.9492	0.9604	0.9618	0.9749	0.9918	0.9975	0.9997
P_R	0.9472	0.9495	0.9601	0.9620	0.9751	0.9918	0.9975	0.9997
$\bar{T}$	13534	13525	13539	13558	13549	13583	13593	13616

Table 12.2 $\hat{P}(\mathrm{CS})$ and sample sizes for experiment 2

Procedure/σ_0^2	0.00	0.01	0.05	0.10	0.25	0.50	0.75	0.90
P_E	0.9469	0.9481	0.9510	0.9495	0.9541	0.9618	0.9707	0.9772
P_R	0.9472	0.9479	0.9501	0.9495	0.9542	0.9616	0.9704	0.9770
$\overline{T}$	13534	13498	13530	13542	13566	13572	13597	13584

all other alternatives. That is, in this setting $\mathrm{Var}(T_{i_1} - T_u) = \mathrm{Var}(T_{i_1}) + \mathrm{Var}(T_u) - 2\mathrm{Cov}(T_{i_1}, T_u)$ and $\mathrm{Cov}(T_{i_1}, T_u) = 0$ while $\mathrm{Var}(T_u)$ is greater with CRN. Hence, $\mathrm{Var}(T_{i_1} - T_u)$ is greater with CRN. But we still achieve greater $\hat{P}(\mathrm{CS})$ because $E(T_u)$ is larger with CRN. The $\hat{P}(\mathrm{CS})$ generally increases as the covariance becomes stronger while the sample sizes are about the same.

12.3.3 *Experiment 3: Best System is Negatively Correlated with Others*

In this experiment, $Z_1 \sim N(0,1)$ and $Z_i \sim N(d^*, 1)$ for $i = 2, 3, \ldots, k$ and correlates with $\mathrm{Cov}(Z_i, Z_j) = \sigma_0^2$, $i \neq j, i \neq 1, j \neq 1$, while $\mathrm{Cov}(Z_1, Z_j) < 0$ for $j = 2, 3, \ldots, k$. Namely, $Z_1 = \xi_1 + \xi_0$. All other settings are the same as previous experiment.

Table 12.3 lists the results. As expected, the $\hat{P}(\mathrm{CS})$s are less than the corresponding $\hat{P}(\mathrm{CS})$s of experiment 2 because system 1 is negatively correlated with all other alternatives, i.e., $\mathrm{Cov}(T_{i_1}, T_u) < 0$. Furthermore, the $\hat{P}(\mathrm{CS})$ generally decrease as the (negative) covariance becomes stronger while the sample sizes are about the same. The negative effect in $\hat{P}(\mathrm{CS})$ of $\mathrm{Cov}(T_{i_1}, T_u) < 0$ is greater than the positive effect of $E(T_u)$ being larger with CRN. Hence, using CRN will backfire if the induced covariances are negative between the best system and all other systems.

12.3.4 *Experiment 4: Unequal Variances*

In this experiment, there are three settings.

- Independent samples: $Z_i \sim N(\mu_i, (2+i)/2)$ for $i = 1, 2, \ldots, k$ and $\mathrm{Cov}(Z_i, Z_j) = 0$, for $i \neq j$.
- Increasing covariance: $Z_i \sim N(\mu_i, (13-i)/2)$ for $i = 1, 2, \ldots, k$ and correlate with

$$\mathrm{Cov}(Z_i, Z_j) = \frac{2}{\sqrt{(13-i)(13-j)}}, \ i \neq j.$$

  That is, $Z_i = \xi_i - \xi_0$ for $i = 1, 2, \ldots, k$, where $\xi_0 \sim N(0,1)$, $\xi_i \sim N(\mu_i, (11-i)/2)$ for $i = 1, 2, \ldots, k$.
- Decreasing covariance: $Z_i \sim N(\mu_i, (2+i)/2)$ for $i = 1, 2, \ldots, k$ and correlate with

$$\mathrm{Cov}(Z_i, Z_j) = \frac{2}{\sqrt{(2+i)(2+j)}}, \ i \neq j.$$

Table 12.3 $\hat{P}$(CS) and sample sizes for experiment 3

Procedure/σ_0^2	0.00	0.01	0.05	0.10	0.25	0.50	0.75	0.90
P_E	0.9469	0.9457	0.9468	0.9435	0.9354	0.9272	0.9252	0.9335
P_R	0.9472	0.9457	0.9462	0.9431	0.9352	0.9272	0.9254	0.9337
$\overline{T}$	13534	13525	13532	13526	13567	13566	13595	13603

Table 12.4 $\hat{P}$(CS) and sample sizes for experiment 4

Procedure/Setting	Independent	Increase	Decrease
P_E	0.9494	0.9749	0.9810
P_R	0.9498	0.9748	0.9811
$\overline{T}$	47337	47470	47366

That is, $Z_i = \xi_i - \xi_0$ for $i = 1, 2, \ldots, k$, where $\xi_0 \sim N(0,1)$, $\xi_i \sim N(\mu_i, i/2)$ for $i = 1, 2, \ldots, k$.

We set $\mu_1 = 0$, $\mu_i = d^*$ for $i = 2, 3, \ldots, k$. All other settings are the same as previous experiment. Hence, the inferior systems have various degrees of correlation with the best system. Table 12.4 lists the results. The $\hat{P}$(CS)s are greater with CRN in both increasing and decreasing covariances cases.

We also performed subset selection using CRN with all the previous configurations, the results are similar.

12.3.5 *Experiment 5: Subset Selection - All Systems are Correlated*

If this experiment, $\xi_0 \sim N(0, \sigma_0^2)$, $\xi_i \sim N(\mu_i, 1 - \sigma_0^2)$ and $Z_i = \xi_i - \xi_0$ for $i = 1, 2, \ldots, k$. We set $\mu_i = 0$, for $i = 1, 2, 3$ and $\mu_i = d^*$, for $i = 4, 5, \ldots, k$. Note that $Z_i \sim N(0, 1)$, for $i = 1, 2, 3$ and $Z_i \sim N(d^*, 1)$ for $i = 3, 4, \ldots, k$ and correlates with

$$\mathrm{Cov}(Z_i, Z_j) = (\frac{1 - \sigma_0^2}{\sigma_0^2} + 1)^{-1/2}(\frac{1 - \sigma_0^2}{\sigma_0^2} + 1)^{-1/2} = \sigma_0^2,\ i \neq j.$$

We set $\sigma_0^2 = 0.00, 0.01, 0.05, 0.10, 0.25, 0.50, 0.75, 0.90$, $P^* = 0.95$, $n_0 = 20$, $k = 10$, and the indifferent amount $d^* = 0.01$. Note that $\xi_0 = 0$ when $\sigma_0^2 = 0.00$, hence, $Z_i = \xi_i$ for $i = 1, 2, \ldots, k$ are independent.

We consider the LFC, $\mu_1 = \mu_2 = \mu_3$, and $\mu_3 + d^* = \mu_4 = \mu_5 = \ldots = \mu_{10}$. The minimum P(CS) should occur at this configuration. Furthermore, 10,000 independent experiments are performed to estimate the actual P(CS) by $\hat{P}$(CS), the proportion of the 10,000 experiments in which we obtained the correct selection.

Table 12.5 lists the results. We list the actual P(CS) of the P_E and P_R procedures with different values of σ_0^2. The $\overline{T}$ column lists the average of the total simulation replications ($\overline{T} = \sum_{r=1}^{10000} \sum_{i=1}^{k} N_{r,i}/10000$, $N_{r,i}$ is the number of total replications or batches for system i in the rth independent run) used in each procedure. As expected, the $\hat{P}$(CS)s increase as the covariance becomes stronger while the average sample sizes are about the

Table 12.5 $\hat{P}$(CS) and sample sizes for experiment 5

Procedure/σ_0^2	0.00	0.01	0.05	0.10	0.25	0.50	0.75	0.90
P_E	0.9529	0.9496	0.9578	0.9587	0.9739	0.9901	0.9966	0.9995
P_R	0.9529	0.9504	0.9579	0.9584	0.9740	0.9900	0.9964	0.9995
$\overline{T}$	8145	8143	8149	8150	8165	8165	8187	8195

same. The $\hat{P}$(CS) of the P_E procedure is slightly less than the nominal value when $\sigma_0^2 = 0.01$. We believe it is because of the stochastic nature of the experiments. Furthermore, the differences of the $\hat{P}$(CS)s of P_E and P_R are very small.

12.3.6 *Experiment 6: Subset Selection - Independence Between Groups 1*

In this experiment, $Z_i \sim N(d^*, 1)$, for $i = 4, 5, \ldots, k$, and $Z_i \sim N(0, 1)$ for $i = 1, 2, 3$ and correlates with $\text{Cov}(Z_i, Z_j) = \sigma_0^2$, $i \neq j, i < 4, j < 4$, while $\text{Cov}(Z_i, Z_j) = 0$ for $i = 1, 2, 3$, and $j = 4, 5, \ldots, k$. All other settings are the same as previous experiment.

Table 12.6 lists the results. As expected, the $\hat{P}$(CS)s are less than the corresponding $\hat{P}$(CS)s of experiment 1 because the best c systems are independent of the other $k - c$ systems. That is, in this setting $\text{Var}(T_c - T_u) = \text{Var}(T_c) + \text{Var}(T_u) - 2\text{Cov}(T_c, T_u)$ and $\text{Cov}(T_c, T_u) = 0$ while $\text{Var}(T_c)$ is greater with CRN. Hence, $\text{Var}(T_c - T_u)$ is greater with CRN. The increased in $\hat{P}$(CS) as the correlation becomes stronger is very small in this setting.

12.3.7 *Experiment 7: Subset Selection - Independence Between Groups 2*

In this experiment, $Z_i \sim N(0, 1)$, for $i = 1, 2, 3$, and $Z_i \sim N(d^*, 1)$ for $i = 4, 5, \ldots, k$ and correlates with $\text{Cov}(Z_i, Z_j) = \sigma_0^2$, $i \neq j, i > 3, j > 3$, while $\text{Cov}(Z_i, Z_j) = 0$ for $i = 1, 2, 3$, and $j = 4, 5, \ldots, k$. All other settings are the same as previous experiment.

Table 12.7 lists the results. Using CRN backfires in this setting. The decrease in $\hat{P}$(CS) becomes larger as the correlation becomes stronger. This result is visualized in Figures 12.5 and 12.6.

12.3.8 *Experiment 8: Subset Selection - Unequal Variances*

In this experiment, there are three settings.

- Independent samples: $Z_i \sim N(\mu_i, (2 + i)/2)$ for $i = 1, 2, \ldots, k$ and $\text{Cov}(Z_i, Z_j) = 0$, for $i \neq j$.
- Increasing covariance: $Z_i \sim N(\mu_i, (13 - i)/2)$ for $i = 1, 2, \ldots, k$ and

Table 12.6 $\hat{P}(\mathrm{CS})$ and sample sizes for Experiment 6

Procedure/σ_0^2	0.00	0.01	0.05	0.10	0.25	0.50	0.75	0.90
P_E	0.9529	0.9534	0.9485	0.9492	0.9499	0.9535	0.9587	0.9644
P_R	0.9529	0.9529	0.9484	0.9495	0.9503	0.9542	0.9587	0.9648
$\overline{T}$	8145	8157	8161	8166	8152	8152	8152	8154

Table 12.7 $\hat{P}(\text{CS})$ and sample sizes for experiment 7

Procedure/σ_0^2	0.00	0.01	0.05	0.10	0.25	0.50	0.75	0.90
P_E	0.9529	0.9502	0.9474	0.9472	0.9408	0.9289	0.9219	0.9161
P_R	0.9529	0.9497	0.9471	0.9466	0.9409	0.9298	0.9230	0.9168
$\overline{T}$	8145	8158	8152	8148	8145	8170	8184	8188

Table 12.8 $\hat{P}$(CS) and sample sizes for experiment 8

Procedure/Setting	Independent	Increase	Decrease
P_E	0.9498	0.9793	0.9665
P_R	0.9500	0.9790	0.9663
$\overline{T}$	28547	28574	28589

correlate with

$$\mathrm{Cov}(Z_i, Z_j) = \frac{2}{\sqrt{(13-i)(13-j)}}, \quad i \neq j.$$

That is, $Z_i = \xi_i - \xi_0$ for $i = 1, 2, \ldots, k$, where $\xi_0 \sim N(0,1)$, $\xi_i \sim N(\mu_i, (11-i)/2)$ for $i = 1, 2, \ldots, k$.

- Decreasing covariance: $Z_i \sim N(\mu_i, (2+i)/2)$ for $i = 1, 2, \ldots, k$ and correlate with

$$\mathrm{Cov}(Z_i, Z_j) = \frac{2}{\sqrt{(2+i)(2+j)}}, \quad i \neq j.$$

That is, $Z_i = \xi_i - \xi_0$ for $i = 1, 2, \ldots, k$, where $\xi_0 \sim N(0,1)$, $\xi_i \sim N(\mu_i, i/2)$ for $i = 1, 2, \ldots, k$.

We set $\mu_1 = 0$, $\mu_i = d^*$ for $i = 2, 3, \ldots, k$. All other settings are the same as previous experiment. Hence, the inferior systems have various degrees of correlation with the best systems.

Table 12.8 lists the results. The $\hat{P}$(CS)s are greater with CRN in both increasing and decreasing covariances cases.

12.4 Summary

We show that it is generally safe to use CRN to increase the probability of correct selection of selection procedures, especially when all systems are correlated with similar strength. The characteristics of the correlated order statistics provides the insight that the increase in $\hat{P}$(CS) is mostly because the expected value of the first-order statistics becomes larger as the (positive) covariances become stronger. The probability of "backfire" decreases when the best systems are correlated with non-best systems. Furthermore, as $P^* < 1$ decreases, it is less likely that the P(CS) will be became smaller as the (positive) covariances become stronger.

Chapter 13

Parallel and Distributed Simulation

Parallel and distributed simulation (PADS) studies how a network of several interconnected models work together to support decision making by distributing the execution of a discrete event simulation (DES) program over multiple computers. Parallel DES programs are executed on multiprocessor computing platforms containing multiple central processing units that interact frequently. Distributed DES programs are executed on loosely coupled systems that may be geographically distributed and require additional interaction times. However, with new computing paradigms such as clusters of workstations and grid computing, the distinction has become un-noticeable. In both cases the execution of a single simulation model, likely composed of several simulation programs, is distributed over multiple processors (computers) and can be executed concurrently. Hence, one can reduce the execution time by up to a factor equal to the number of processors that are used. Distributing the execution across multiple computers and utilizing the resources of many computer systems is also beneficial in allowing the execution of larger simulations where the capacity of one computer may not be enough to carry out the simulation. A more detailed discussion of distributed and parallel simulation can be found in [Fujimoto (2000)].

13.1 Introduction

A logical process (LP) is a distinct flow of control, containing a combination of computation and operation. The simulation of each design can be treated as a LP and selection involves simulating a collection of LPs. Since system designs are independent of each other, the simulation of each system, i.e, LP, can be performed independently in a parallel and distributed fashion.

The main goal is to compute the results of the simulation as quickly as possible to improve the effectiveness of the simulation tool. Thus, our immediate concern is capability rather than run-time performance. If a better alternative is found early in the process, it can be used to eliminate inferior designs at an early stage during the simulation process. When all alternatives are divided into several non-overlapping groups, if the best sample mean of all alternatives at any given moment is used to eliminate only those inferior designs within the same group, the overall efficiency may suffer. However, we will be able to process R&S of several groups in parallel, i.e., the entire R&S can be performed by a set of concurrently executing processes. Thus, the duration of run time will be decreased. Furthermore, it is possible to compute the best sample mean of all alternative designs from all groups at any iteration and use the best sample mean to eliminate inferior designs in all groups. Even though the application of parallel simulation technology has been limited, there has been some work done in this area, for example, [Luo et al. (2000)] deploy OCBA (Optimal Computation Budget Allocation) to distribute simulation replications over the web for R&S problems.

13.2 Parallel and Distributed Selection

In this section, we present the basis of selection in a parallel and distributed fashion. PADS attempts to decrease simulation analysis time by distributing the simulation workload among multiple computers (processors). A simulation program operates on a models state variables by executing a time-ordered sequence of simulation events. Each event may change the state of the simulated system and schedule one or more future events. In most discrete-event simulation, the order in which events are executed is stored in the event list and is determined by a next-event time advance mechanism for the simulation clock, see [Law (2014)]. Events are executed in nondecreasing time-stamp order so that the simulation clock always advances. A conventional PADS decomposes a simulation model into communicating LPs to perform different events. The PADS procedure maps each LP to a processor and uses interprocessor communication to allow LP on different processors to communicate with each other.

The application of parallel and distributed simulation has been limited. It is generally difficult to implement simulation in a parallel and distributed environment because of the sequential nature of most simulations. However, simulation based ranking and selection is well suited for parallel and distributed simulation because the behavior of each system can be simulated independently. For example, a network of computers can be used to perform R&S problems, several computers can be assigned to simulate the performance of one or several systems, while a computer dedicated to perform the comparisons between systems orchestrates the overall simulation strategy.

To set up the procedure, let $G_1, G_2, \ldots, G_m$ be groups of designs such that $G_1 \cup G_2 \cup \ldots \cup G_m = \{1, 2, ..., k\}$, $G_i \cap G_j = \emptyset$ for $i \neq j$, $k_i = |G_i| \geq 2$ for all i, where $|I|$ denotes the cardinality of the set I. When we perform selection in group e, the designs in G_e will be compared to others in the same group. Without loss of generality, assume $i_1 \in G_g$ for some $1 \leq g \leq m$. For each group e, let $\bar{X}_b^e = \min_{i \in G_e} \bar{X}_i$ denote the best sample mean in group e. In each group, designs $q \in G_e$ such that $\bar{X}_q \leq \bar{X}_b^e + d^*$ will be considered as surviving designs. If we perform the selection in each group e, then $P[i_1$ surviving G_g selection$] \geq P$. Since the procedure has allocated substantial sample sizes to those surviving designs, the additional samples in subsequent selections will most likely be small. The surviving designs are then grouped into a single group or multiple groups depending on the number of surviving designs. Note that the probability that design i_1 is among the set of surviving designs is at least P. To simplify our discussion, we group the surviving designs into a single group. Let k' denote the number of surviving designs and let the set I contain the k' surviving designs. We then select from all surviving designs in I. Note that for each design $i \in I$, we have the triple $(n_i, \sum_{j=1}^{n_i} X_{ij}, \sum_{j=1}^{n_i} X_{ij}^2)$. We perform all pairwise comparisons among these k' designs to eliminate inferior designs, i.e., design j having $\bar{X}_j > \bar{X}_i + w_{ij}$ for some $i \in I$ will be excluded from further simulation. Furthermore, if $w_{ij} < d^*$ and $\bar{X}_i > \bar{X}_j$ for $i, j \in I$, remove i from I. Let r be the iteration index and $\hat{d}_i = \max(d^*, \bar{X}_i - U(\bar{X}_b))$ for $i \in I$, where $\bar{X}_b = \min_{i \in I} \bar{X}_i$. Compute $\delta_{i,r} = \lceil (h_t S_i(n_i)/\hat{d}_i)^2 \rceil - n_i$. Let $\hat{\delta_r} = \min_{i \in I} \delta_{i,r}$. Simulate additional δ_r replications or batches for each design $i \in I$ at iteration r. If $\mu_{i_2} - \mu_{i_1} > d^*$, the probability of i_1 being eliminated by some design $i \neq i_1$ is less than $(1-P^*)/(k-1)$. Hence, by the Bonferroni inequality, the probability that design i_1 survives the selection is larger than $P^* = 1 - \sum_{i=1}^{k-1}(1 - P^*)/(k-1)$.

We can increase the efficiency of deploying the selection procedure in a parallel and distributed environment by passing the best sample mean at each iteration to all groups. Let $\bar{X}_B$ be the best sample mean from all groups, i.e., $\bar{X}_B = \min_{1 \leq e \leq m} \bar{X}_b^e$, and pass the triple

$(n_B, \sum_{j=1}^{n_B} X_{Bj}, \sum_{j=1}^{n_B} X_{Bj}^2)$ to all groups. Since $\bar{X}_B \leq \bar{X}_b^e$ for $1 \leq e \leq m$, the overall efficiency of selection may be improved. Note that $\bar{X}_b^e$ from different groups may be obtained at different iteration with different sample sizes.

Another benefit of distributed simulation is the ability to integrate several different simulators, i.e., different simulation software packages, into a single simulation environment. Since the underlying simulator used to generate the simulation results has no impact on the final selection, we can simulate alternative system designs with different simulation packages. The ability to integrate different simulators allows the simulation models to be executed and developed concurrently.

Sequential selection procedures offer a natural way of performing selection in a parallel and distributed fashion, providing further gains in simulation efficiency. Once the number of alternative designs and number of groups have been determined, several control computers, denoted Ω, can be used to orchestrate the execution of simulation. These Ω computers will initiate the execution of simulation and invoke a set of remote computers, for example through remote method invocation, to generate r samples of one or several system designs. That is, each system is evaluated through distributed runs of discrete simulation models. The simulation results can be written to a shared file server or ftp (file transfer protocol) to the computers Ω that are performing the R&S. Based on the analysis, computers Ω send the required additional sample size for each system design to the corresponding computer. The communication between these computers will repeat iteratively until simulation is complete. Deploying selection procedures in parallel and distributed environment requires little communication between processors and overhead, making it practical and attractive.

13.3 The Framework

The use of networks of workstations interconnected through LAN/WANs (Local Area Network/Wide Area Network) has evolved into an effective and popular platform for parallel and distributed computing. The advantages of these network computing environments include 1) ready availability, 2) low cost, and 3) incremental scalability. Furthermore, network computing environments retain their ability to serve as a general-purpose computing platform and to run commercially available software products. One difficulty associated with PADS is time management when ensuring that the execution of the distributed simulation, i.e., LP, is properly synchronized. Communication must be sent between processors when corresponding parts of the model interact logically. As a result, issues concerning the sub models ability to proceed at its own pace arise because generally PADS LP can schedule further events not only for itself but also for other LPs. Therefore, events cannot be executed in a straightforward time-stamp manner. As pointed out earlier, for the R&S procedure to perform correctly, it only needs the triple $(n_i, \sum_{j=1}^{n_i} X_{ij}, \sum_{j=1}^{n_i} X_{ij}^2)$ from each design. Thus, time management does not complicate processing R&S in parallel. In this case, each LP is viewed as an independent and autonomous discrete event simulator. This means that each LP maintains its local state information corresponding to the entities it is simulating and a list of events that have been scheduled for this LP. Furthermore, each LP only schedules additional events for itself but not for other LPs. Hence, it is straightforward to implement R&S in a parallel and distributed fashion.

The World Wide Web or the Internet is a loose coupling of thousands of networks and millions of computers around the globe. The Internet has become one of the most important information sources and communication platforms in industry. An inherent characteristic of the Internet is its distributed nature, hence, it provides an excellent basis for distributed simulation.

13.4 Selection with All Pairwise Comparisons

Most selection procedure requires the input data are i.i.d. (independent and identically distributed) normal. Many performance measures of interest are taken over some average of a sample path or a batch of samples. Thus, many applications tend to have a normally distributed simulation output. If the non-normality of the samples is a concern, users can use batch means (see [Law (2014)]) to obtain samples that are essentially i.i.d. normal.

Effective reduction of computation efforts while obtaining a good decision is crucial. It has been proposed to sequentialize selection procedures to eliminate the drawback of two-stage procedures and to improve its efficiency. [Rinott (1978)] procedure and its variants are derived based on P(CS)=$\Pr[\bar{X}_{i_1} < \bar{X}_{i_l}$, for $l = 2, 3, \ldots, k] \geq P^*$. To further improve the efficiency of sequentialized selection procedure we incorporate all pairwise comparisons at each iteration. Let $P = 1-(1-P^*)/(k-1)$. Inferior design i such that $\Pr[\bar{X}_i > \bar{X}_j] \geq P$ of some design j will be excluded from further simulation at each iteration.

For completeness, the related theorem and proposition are listed here.

Theorem 13.1. *For k competing designs whose performance measure X_{ij} are normally distributed with means $\mu_1, \mu_2, \ldots, \mu_k$ and unknown variances that need to be estimated by sample variances $S_1^2(n_c), S_2^2(n_c), \ldots, S_k^2(n_c)$, where n_c is the current sample size, P(CS) will be at least P^* when the sample size for design i is*

$$N_i = \max(n_c, \lceil (h_t S_i(n_c)/d_i)^2 \rceil), \; for \; i = 1, 2, \ldots, k,$$

where n_c is the current sample size, the critical value $h_t = \sqrt{2} t_{n_c-1,P}$, $P = 1 - (1 - p^*)/(k-1)$, and $d_i = \max(d^*, \mu_i - \mu_{i_1})$.

Let the set I contain competing designs and let $\bar{X}_b = \min_{i \in I} \bar{X}_i$ at each iteration. Since the true means μ_i are unknown, d_i is conservatively estimated by $\hat{d}_i = \max(d^*, \bar{X}_i - U(\bar{X}_b))$, where $U(\bar{X}_b)$ is the upper one-tailed P^* confidence limit of μ_b, i.e., $\Pr[\mu_b \leq U(\bar{X}_b)] \geq P^*$. The value of $t_{n_c-1,P}$ can be approximated easily, see Section 1.5.3 and [Hastings (1955)].

Theorem 13.2. *Let the set I contain k competing designs whose performance measure are normally distributed with unknown means and unknown variances that need to be estimated by sample means $\bar{X}_1, \bar{X}_2, \ldots, \bar{X}_k$, and sample variances $S_1^2(n_1), S_2^2(n_2), \ldots, S_k^2(n_k)$. If $k-1$ designs are removed (eliminated) sequentially from I with each eliminated design j satisfies the equation that $\bar{X}_j(n_c) > \bar{X}_i(n_c) + w_{ij}, \quad i, j \in I$, where n_c is the current sample size for each design, $w_{ij} = t_{n_c-1,P}\sqrt{S_i^2(n_c)/n_c + S_j^2(n_c)/n_c}$ and $P = 1 - (1 - p^*)/(k-1)$, then $Pr[i_1 \in I] \geq P^*$.*

The Sequential Selection Procedure with All Pairwise Comparisons (SAPC):

(1) Initialize the set I to include all k designs. Let $N_{i,r}$ and $\bar{X}_{i,r}$, respectively, be the allocated sample size and the sample mean of design i at the rth iteration. Simulate n_0 replications or batches for each design $i \in I$. Set the iteration number $r = 0$, and $n_c = N_{1,r} = N_{2,r} = \cdots = N_{k,r} = n_0$, Set $P = 1 - (1 - P^*)/(k-1)$.
(2) Perform all pairwise comparisons and delete inferior design j from I; i.e., $\bar{X}_j > \bar{X}_i + w_{ij}$, $i, j \in I$. Note w_{ij} is the one-tailed P c.i. half-width.
(3) If $w_{ij} < d^*$ and $\bar{X}_j > \bar{X}_i$, remove design j from I.

(4) If there is no more than one element (or the pre-determined number of best designs) in I, go to step 8.

(5) Compute the critical value $h_t = \sqrt{2}t_{n_c-1,P}$.

(6) Let $\bar{X}_{b,r} = \min_{i\in I} \bar{X}_{i,r}$, For all $i \in I$, compute $\hat{d}_{i,r} = \max(d^*, \bar{X}_{i,r} - U(\bar{X}_{b,r}))$, where $U(\bar{X}_{b,r})$ be the upper one-tailed P^* confidence limit of μ_b at the rth iteration, and compute

$$\delta_{i,r+1} = \lceil ((h_t S_i(n_c)/\hat{d}_{i,r})^2 - n_c)^+ \rceil.$$

(7) Set $r = r + 1$. If $\delta_{i,r} = 0$, set $\delta_{i,r} = 1$. Set the incremental sample size at the rth iteration $\delta_r = \min_{i\in I} \delta_{i,r}$. For $\forall i \in I$, simulate additional δ_r samples, set $n_c = n_c + \delta_r$. Go to step 2.

(8) Return the values b and $\bar{X}_b(N_b)$, where $\bar{X}_b(N_b) = \min \bar{X}_i(N_i)$, $1 \le i \le k$ and i was not eliminated by all pairwise comparisons.

In the sequential selection procedure, all the alternatives $1 \le i \le k$ are initially included in the set I for R&S. If all $k-1$ designs were eliminated from I through the two-sample-t test, then $\Pr[i_1 \in I] \ge P^*$. On the other hand, if some designs were eliminated from I because $w_{ij} < d^*$, then the procedure can only guarantee P(CS) $\ge P^*$. The basic idea is that the procedure sequentially removes $k-1$ designs from I. If $\mu_{i_2} - \mu_{i_1} \ge d^*$, then the probability of wrongly removing design i_1 is $1-P$ each time a design is removed. By the *Bonferroni* inequality $\Pr[i_1 \in I] \ge 1 - \sum_{i=1}^{k-1}(1-P) = P^*$. We use the equation $S_i^2(n_c) = (\sum_j^{n_c} X_{ij}^2/n_c - \bar{X}_i^2)n_c/(n_c-1)$ to compute the variance estimator so that we are only required to store the triple $(n_c, \sum_{j=1}^{n_c} X_{ij}, \sum_j^{n_c} X_{ij}^2)$, instead of the entire sequences $(X_{i1}, X_{i2}, \ldots, X_{in_c})$.

13.5 Empirical Experiments

In this section we list the empirical results of SAPC and running SAPC under a parallel and distributed environment, denoted NPDS. Furthermore, NPDS$_s$ denotes performing selection separately within each group, i.e., the best sample mean from other groups is not used to eliminate inferior in current group. Instead of using stochastic systems simulation examples, which offer less control over the factors that affect the performance of a procedure, we use various normally distributed random variables to represent the system performance measures.

In the experiment, there are 10 alternative designs under consideration, and each $X_{ij} \sim N(\mu_i, (\sqrt{10})^2)$, where $N(\mu, \sigma^2)$ denotes the normal distri-

Table 13.1 $\hat{P}$(CS) and sample sizes

	$P^* = 0.90$		$P^* = 0.95$	
Procedure	$\hat{P}$(CS)	$\overline{T}$	$\hat{P}$(CS)	$\overline{T}$
SAPC	0.9567	535	0.9801	746
NPDS	0.9458	531	0.9760	777
NPDS$_s$	0.9561	659	0.9744	842

Table 13.2 Detailed sample sizes of NPDS

Design	μ	Sample	Within	Sample	Select
1	0	65	0.9737	96	0.9458
2	1	62	0.4156	73	0.0113
3	1	62	0.4174	73	0.0116
4	1	62	0.4215	73	0.0119
5	2	52	0.0093	52	0.0
6	1	34	0.4342	50	0.0181
7	2	29	0.0265	29	0.0004
8	2	28	0.0294	29	0.0003
9	2	28	0.0311	29	0.0006
10	3	23	0.0005	23	0.0

bution with mean μ_i and variance σ^2. See Table 13.2 for the values of μ_i. We divide these designs into two non-overlapping groups: group 1 contains designs 1 through 5, group 2 contains designs 6 though 10. The indifference amount d^* is set to 1.0 in all cases. We set the initial replication $n_0 = 10$ and the minimal probability of CS $P^* = 0.90$ and $P^* = 0.95$. Furthermore, 10,000 independent experiments are performed to estimate the actual P(CS) by $\hat{P}$(CS): the proportion of the 10,000 experiments in which we obtained the correct selection: design 1.

Table 13.1 lists the experimental results. The SAPC, NPDS, and NPDS$_s$ rows list the results of each procedure. The $\hat{P}$(CS) column lists the proportion of correct selection. The $\overline{T}$ column lists the average of the total simulation replications ($\overline{T} = \sum_{r=1}^{10000} \sum_{i=1}^{k} N_{r,i}/10000$, $N_{r,i}$ is the total number of replications or batches for design i in the rth independent run) used in each procedure. The observed P(CS)'s are greater than the specified nominal levels of 0.90 and 0.95. The sample sizes allocated by NPDS are about the same as those allocated by SAPC since the best design is used to eliminate inferior designs early in the selection process in all groups. However, NPDS can perform selection in parallel and may result in a shorter run time, especially when the best sample means in each group is about the same.

Table 13.2 lists the results when the best sample mean is available to all

Table 13.3 Detailed sample sizes of $NPDS_s$

Design	μ	Sample	Within	Sample	Select
1	0	90	0.9934	101	0.9561
2	1	66	0.3613	67	0.0095
3	1	67	0.3588	67	0.0074
4	1	66	0.3552	67	0.0084
5	2	26	0.0120	26	0.0
6	1	91	0.9930	100	0.0184
7	2	67	0.3624	67	0.0
8	2	67	0.3633	67	0.0001
9	2	67	0.3645	67	0.0001
10	3	26	0.0112	26	0.0

groups. The μ column lists the true mean of design i. The Within column lists the proportion that $\bar{X}_i \leq \bar{X}_b + d^*$, where $\bar{X}_b = \min_{1 \leq i \leq k} \bar{X}_i$ is the best sample mean of all groups. The Select column lists the proportion that the particular design has the best sample mean. The first Sample column lists the resulting sample sizes from the subdivided groups. The second Sample column lists the final sample sizes.

Table 13.3 lists the results when the best sample mean is available only to the group contain this particular design. The Within column under NPDS without sharing list the proportion that particular design has sample mean within the best sample mean in the same group (instead of the best sample mean of all groups) plus the indifference amount. All other fields are as defined in Table 13.2. The detailed results for $P^* = 0.95$ are similar. The allocated sample sizes for designs in the second group, i.e., designs 6 through 10, are significantly reduced when the best sample means from all groups are used to eliminate inferior designs. Even though the means within each group have similar configuration, the resulting sample sizes and within from each group are not similar. This is because design 6 is no longer the best design in group 2 when the best sample from all groups is included for selection.

13.6 Summary

Parallel and distributed simulation can reduce execution time for time-consuming applications, such as ranking and selection of stochastic systems. We have presented a framework for deploying selection procedures in a parallel and distributed environment. All the alternative designs are subdivided into several groups and the entire selection is performed by a set of concurrently executing processes. Thus, we may be able to shorten the run time. The procedure incorporates all pairwise comparisons to eliminate inferior designs at each iteration, which may reduce the overall computational effort. The proposed procedure takes into account the difference of sample means when determining the sample sizes and is suitable even when the number of designs under consideration is large.

Chapter 14

Multi-Objective Selection

The ranking and selection studies in previous chapters have focused on selection based on a single measure of system performance. In many practical situations, however, we need to select systems based on multiple criteria (attributes, objectives, or performance measures). For example, in product-design optimization, the cost and the quality of products are two conflicting objectives. In evaluating airline flight schedules, we may want to select flight schedules in terms of minimal (flight and ground) staffs and minimal percentage of late arrivals. In this setting, the problem of selecting the best systems from a set of alternatives through simulation becomes a multi-objective ranking and selection problem. Researchers have developed procedures to address this problem. [Butler et al. (2010)] combine the multiple attribute utility theory with the [Rinott (1978)] procedure to handle multiple performance measures. Before performing selection, users transform multiple performance measures into one utility score. The procedure then finds the system that gives the highest utility. [Dudewicz and Taneja (1978)] propose a multivariate procedure by defining a multivariate normal vector composed of $\omega > 1$ component variates with an unknown and unequal variance-covariance matrix. They redefine the indifference-zone parameter as the Euclidean distance from a mean vector to the best mean vector.

[Lee et al. (2004)] point out that when there are multiple performance measures the selected "best" system would be strongly dependent on the decision maker's preference. Instead, they develop a multi-objective computing budget allocation (MOCBA) procedure which incorporates the concept of Pareto optimality into the ranking and selection scheme to find all non-dominated solutions, i.e., the non-dominated Pareto set . In the case that the problems are multi-objective in nature, there may not exist a single best solution, but rather a set of non-dominated solutions is referred to as the pareto set of solutions. They represent the "best" systems and are characterized by the definition that no other solution exists that is superior in all the objectives. A solution is called Pareto-optimal if there exists no other solution that is better in all criteria. In general, eventually a single system must be selected for a given situation. In the planning phase, however, it is desirable to have a set of good options available. For example, a company could manage several different warehouses, each with its own requirements. Thus, each warehouse can select a final system from the set of good options according to its own requirements. Moreover, many mail-order companies offer several different delivery options of their products: 1) the shortest delivery time; 2) the least expensive; 3) some other delivery options with modest delivery time and modest cost. Each options are non-dominated. Customers can choose the delivery option based on their own need. The MOCBA procedure aims to determine the Pareto set by allocating the sample sizes intelligently and minimize the required sample sizes.

[Lee et al. (2006)] integrate selection procedures with search mechanism to solve multi-objective simulation-optimization problems. In the application of using evolutionary algorithms to solve multi-objective problems, the concept of Pareto optimality is often employed to find the non-dominated Pareto set. In this chapter, we extend a R&S procedure to select a Pareto set containing non-dominated systems. We try to provide a non-dominated Pareto set of systems to the decision maker, rather than reducing the problem to a single-objective model and providing a single "best" system. We attempt to minimize the required sample sizes that provide a probability guarantee that the selected systems are non-dominated. Alternatively, the procedure can be used to maximize the probability of correctly selecting non-dominated systems given a fixed sample size.

14.1 Introduction

In this section, we introduce the necessary notation and background:

- ω: the number of performance measures of interest,
- q: the index of performance measures, i.e., $q = 1, 2, \cdots, \omega$,
- X_{iqj}: the independent and normally distributed observations from the jth replication or batch of the qth performance measure of the ith system,

N_i: the total number of replications or batches for system i,

n_i: the intermediate number of replications or batches for system i,

μ_{iq}: the qth expected performance measures of system i,

$\vec{\mu}_i$: the vector of ω expected performance measures of system i, i.e., $\vec{\mu}_i = (\mu_{i1}, \mu_{i2}, \ldots, \mu_{i\omega}) = (E(X_{i1j}), E(X_{i2j}), \ldots, E(X_{i\omega j}))$,

$\bar{X}_{iq}$: the sample mean of the qth performance measure of system i with n_i samples, i.e., $\sum_{j=1}^{n_i} X_{iqj}/n_i$,

σ_{iq}^2: the variance of the qth observed performance measure of system i from one replication or batch, i.e., $\sigma_{iq}^2 = \text{Var}(X_{iqj})$,

$S_{iq}^2(n_i)$: the sample variance of the qth performance measure of system i with n_i replications or batches, i.e., $S_{iq}^2(n_i) = \sum_{j=1}^{n_i}(X_{iqj} - \bar{X}_{iq})^2/(n_i - 1)$.

14.1.1 *A Multi-Objective Selection Procedure*

Based on the Bayesian methodologies, [Lee et al. (2004)] develop a procedure to select the Pareto set based on a performance index ψ_i that measures the sum of the probabilities that other systems are better than a given system i. Let $P(\vec{\mu}_l \prec \vec{\mu}_i)$ denote the probability that system l dominates system i. Then

$$\psi_i = \sum_{l=1, l\neq i}^{k} P(\vec{\mu}_l \prec \vec{\mu}_i) = \sum_{l=1, l\neq i}^{k} \prod_{q=1}^{\omega} P(\mu_{lq} \leq \mu_{iq}),$$

and at least one of those inequalities is strict. Note that ψ_i is not a probability, it is a function of probability and can be greater than 1. The procedure requires a user-specified parameter ψ^*, a performance index for systems in the Pareto set to be retained at the end of the simulation. Furthermore, the procedure assumes that the number of non-dominated systems is known in advance. The procedure does not incorporate the indifference-zone approach, it performs an optimization to minimize sample sizes subject to the constraints that $\psi_i \leq \psi^*$ for all system i in the Pareto set. Alternatively, the procedure can optimize $\sum \psi_i$ subject to the constraints that the total sample size is less than a user specified computing budget T. The parameter ψ^* is analogous to the parameter P^* in the procedure, but ψ^* is not as easy to specify or interpret.

14.2 Methodologies

In this section we present a strategy of applying R&S technique to select a Pareto set of non-dominated systems. We assume that the performance measures are independent from one another. Like other selection procedures, the proposed procedures assume input data are i.i.d. normal and allow unknown and unequal variances across systems. If non-normality of the input data is a concern, users can use batch means to obtain sample means that are essentially i.i.d. normal.

14.2.1 *Prolog*

The problem considered in this study is as follows. Suppose that we have a set of k systems, where each is evaluated in terms of ω independent objectives. We want to find the non-dominated (Pareto) set of systems by running simulations. The goal is to determine an allocation of the simulation replications to the systems, so that there is certain probability guarantee that the selected systems are true non-dominated.

We apply the SRS selection procedure in each objective (performance measure) to select $m_p < k$ non-dominated systems. The selected systems are non-dominated because they are the best in at least one objective. The collection of these non-dominated systems is an incomplete Pareto set, because the collection may not include all the non-dominated systems. To simplify the discussion, we assume $m_p \leq \omega$. That is, we select no more than one system in each objective initially. The procedure obtains approximately P^* probability that the selected systems are the best in at least one performance measure and, thus, are non-dominated. Here the value of m_p and P^* are both specified by the users.

To adapt the procedure for not using the indifference amount d^*, the controlled distances will be computed as follows.

$$d_{b_l} = \begin{cases} \bar{X}_{b_2} - \bar{X}_{b_l} & l = 1 \\ \bar{X}_{b_l} - \bar{X}_{b_1} & l = 2, 3, \ldots, k. \end{cases} \tag{14.1}$$

Then the required sample size is computed by

$$N_{b_l} = \max(n_0 + 1, \lceil (hS_{b_l}(n_0))/d_{b_l})^2 \rceil) \text{ for } l = 1, 2, \ldots, k. \tag{14.2}$$

Furthermore, the incremental sample size

$$\delta_{i,r+1} = \begin{cases} (N_{i,r+1} - N_{i,r})^+ & \text{if } N_{i,r+1} - N_{i,r} < 5 \\ 5 & \text{otherwise.} \end{cases} \tag{14.3}$$

Table 14.1 Means and standard deviations of systems

Systems	Mean$_1$	Std$_1$	Mean$_2$	Std$_2$	...	Mean$_\omega$	Std$_\omega$
1	μ_{11}	σ_{11}	μ_{12}	σ_{12}	...	$\mu_{1\omega}$	$\sigma_{1\omega}$
2	μ_{21}	σ_{21}	μ_{22}	σ_{22}	...	$\mu_{2\omega}$	$\sigma_{2\omega}$
⋮	⋮	⋮	⋮	⋮	⋮	⋮	⋮
k	μ_{k1}	σ_{k1}	μ_{k2}	σ_{k2}	...	$\mu_{k\omega}$	$\sigma_{k\omega}$

Here r is the iteration index.

This is because the difference of the best sample mean and the second best sample mean could be very small and without a user specified d^* as a lower bound, the estimated sample size could be much larger than necessary. The choice to limit the incremental sample size to 5 is somewhat arbitrary, however, we believe it provides a good compromise between the number of iterations and the total sample size.

14.2.2 *The Strategy*

To facilitate the discussion, we assume that the mean and standard deviation of the ω performance measure of k systems are as listed in Table 14.1. Let $T_q = \sum_i^k v_{iq}$ be the minimal required sample size to correctly select the system having the best performance measure q, where v_{iq} is the optimal sample size for system i based on performance measure q. Without loss of generality, assume $T_1 \le T_2 \le \ldots \le T_\omega$. Hence, the minimal required sample size to obtain m_p non-dominated systems should be $\sum_{i=1}^k \max_{q=1}^{m_p}(v_{iq}) \ge T_{m_p}$. We denote this the theoretical optimal sample sizes allocation of selecting a Pareto set TOAP. Note that $\sum_{i=1}^k \max_{q=1}^{m_p}(v_{iq}) = T_{m_p}$ when $v_{im_p} = \max_{q=1}^{m_p}(v_{iq})$ for $i = 1, 2, \ldots, k$. The procedure simulates n_0 samples initially and computes the sample mean $\bar{X}_{iq}$ and sample variance $S_{iq}^2(n_i)$ for $i = 1, 2, \ldots, k$ and $q = 1, 2, \ldots, \omega$. For $q = 1, 2, \ldots, \omega$, it ranks the sample means such that $\bar{X}_{b_1q} \le \bar{X}_{b_2q} \le \ldots \le \bar{X}_{b_kq}$ and computes the t score

$$T_q = \frac{\bar{X}_{b_2q} - \bar{X}_{b_1q}}{\sqrt{S_{b_2q}^2(n_{b_2})/n_{b_2} + S_{b_1q}^2(n_{b_1})/n_{b_1}}}.$$

It then ranks the t scores T_q for $q = 1, 2, \ldots, \omega$ such that $T_{q_1} \le T_{q_2} \le \ldots \le T_{q_\omega}$. Let $P = (P^*)^{1/(k-1)}$, $n_{b_s} = \min(n_{b_1}, n_{b_2})$. If $T_q > t_{n_{b_s}-1,P}$, then we declare that system b_1 is non-dominated with respect to performance measure q. The procedure terminates when it finds the non-dominated systems under m_p performance measures or the incremental sample size of system i $\delta_i = 0$ for $i = 1, 2, \ldots, k$. Note that a system can be declared non-dominated under more than one performance measure.

In each iteration, the procedure finds the largest T_q that is less than $t_{n_{b_s}-1,P}$, i.e., a non-dominated system has not been declared and a system is the most likely to be declared non-dominated with respect to performance measure q than other performance measures. Hence, the procedure allocates more samples according to the estimated incremental sample size of performance measure q. This approach is similar to the greedy algorithm of linear programming.

14.2.3 *The Incomplete Pareto Set Selection Procedure*

We denote the extended procedure: the SRSIP (Sequential Ranking and Selection of an Incomplete Pareto Set) procedure.

Remark: Specify $m_p \leq \omega$, the number of performance measures that non-dominated systems will be selected and n_0 the number of initial sample size. Initialize the counter of the number of systems that has been declared non-dominated $l = 0$.

The SRSIP Procedure:

(1) For each performance measure $q = 1, 2, \ldots, \omega$ perform steps 2 and 3 of the SRS procedure.
(2) Rank the t scores such that $T_{q_1} \leq T_{q_2} \leq \ldots \leq T_{q_\omega}$.
(3) For each q such that $T_q > t_{n_{bs}-1,P}$, we declare that system b_1 is non-dominated with respect to performance measure q and set $l = l + 1$.
(4) Find the performance measure q_l such that $T_{q_l} = \max(T_q | T_q < t_{n_{bs}-1,P}, q = 1, 2, \ldots, \omega)$. Note that the degrees of freedom $n_{bs} - 1$ may be different between different performance measures.
(5) For each performance measure q such that $N_{i,r+1} \leq N_{i,r}$, for $i = 1, 2, \ldots, k$, then set $l = l + 1$.
(6) If $l \geq m_p$, then go to step 8.
(7) If $N_{i,r+1} - N_{i,r} < 5$, set $\delta_{i,r+1} = (N_{i,r+1} - N_{i,r})^+$. Otherwise, set $\delta_{i,r+1} = 5$. Simulate $\delta_{i,r+1}$ additional samples for system i. Set $N_{i,r+1} = N_{i,r} + \delta_{i,r+1}$ and $r = r + 1$. Go to step 1.
(8) Return the best system of each performance measures q_l, for $l = 1, 2, \ldots, m_p$.

The procedure can easily be modified when the total sample size is fixed. Furthermore, the stopping conditions can be revised so that the procedure terminates when non-dominated systems are declared from m_p performance measures and/or non-dominated systems are declared under certain performance measures. Let ξ_i denote the probability that the selected system i is non-dominated. The SRSIP procedure guarantees that the selected systems i are non-dominated with $\xi_i \geq P^*$; but there is no guarantee that all non-dominated systems are selected. Conversely, there is no more $1 - P^*$ probability that a selected system is dominated. Let M_p denote the current (incomplete) Pareto set. If the size of M_p is $I = |M_p|$, then the probability that the selected Pareto set contains a dominated system is less than $(1 - P^*)I$.

14.2.4 *The Two-Stage Pareto Set Selection Procedure*

Once those non-dominated systems (that are the best in at least one objective) are selected, a subsequent process can be performed to select all non-dominated systems. Let M_p^C denote the complement of M_p, i.e., the set of systems under consideration j such that $j \notin M_p$. For each system $j' \in M_p^C$, we will compare it against every system $i \in M_p$ to find other non-dominated systems. This process is carried out by setting each $j' \in M_p^C$ as non-dominated and is (temporarily) designated as non-dominated when no design $i \in M_p$ is found to dominate it. Note that M_p is updated sequentially, i.e., each time a non-dominated design is found. The procedure then performs all-pairwise comparisons among systems that are (temporarily) designated as non-dominated to remove systems that are dominated.

In the best scenarios that no more than one system is (temporarily) designated as non-dominated, no pairwise comparisons is required. In the worst scenarios that all systems that are not in the Pareto set are (temporarily) designated as non-dominated, with the comparisons already performed the second-phase has performed all-pairwise comparisons among the k systems. However, with the sample sizes allocation strategy described in Section 14.2.3 there is no statistical guarantee that the non-dominated systems that are not the best in any objectives will be included in the Pareto set. The reason is that under some performance measure q the best system may be much better than all other systems, i.e., $\mu_{i_2 q} - \mu_{i_1 q} >> \mu_{i_a q} - \mu_{i_b q}$ where $a > b$ and $a, b \neq 1$. Note that $\mu_{i_1 q} \leq \mu_{i_2 q} \leq \ldots \leq \mu_{i_k q}$. Consequently, relatively smaller sample sizes are enough to select the best system with the required precision; but the allocated sample sizes are not large enough to rank any other two systems with the required precision. Note that that the smaller the difference of $\mu_{i_a q} - \mu_{i_b q}$ the larger the required sample sizes to compare these two systems correctly.

14.2.5 *Incorporating Indifference-Zone*

A promising solution is to incorporate the indifference-zone approach into selecting the Pareto set. Let d_q be the user specified indifference amount of performance measure q. For $i = 1, \ldots, k$ and $q = 1, 2, \ldots, \omega$, let the required sample size of design i based on performance measure q

$$N_{iq} = \max(n_0 + 1, \lceil (hS_{iq}(n_0))/d_q)^2 \rceil). \tag{14.4}$$

Then the required sample size for design i will be

$$N_i = \max_{q=1}^{\omega} N_{iq}. \tag{14.5}$$

The procedure then allocates all $N_i - n_0$ samples for design i in the second stage. The selection process is developed as a two-stage-Pareto-set-selection procedure (TSP), i.e., the procedure first selects non-dominated systems that are the best in some performance measure, then selects other non-dominated systems that are not the best in any performance measures. Figure 14.1 lists a high-level flowchart of TSP.

The rationale is that we are performing several pairwise comparisons among systems in each performance measure. Note that if $0 < d_{abq} = \mu_{aq} - \mu_{bq}$, then $P_{abq} = \Pr[\bar{X}_{bq} < \bar{X}_{aq}] \geq 0.5$. The value of P_{abq} increases as the value of d_{abq} increases. With the sample sizes N_i, if the true performance measure q of designs are deviated more than d_q, then the procedure can rank any two systems correctly with high confidence, e.g., $\Pr[\bar{X}_{bq} < \bar{X}_{aq}] \geq P = (P^*)^{1/(k-1)}$ when $\mu_{bq} + d_q \leq \mu_{aq}$. Consequently, we will have P^* confidence that the selected best systems are non-dominated. Similarly, when selecting other designs that belong to the Pareto set but are not the best in any performance measures, there is P probability that the selected design belongs to the Pareto set when for each incumbent non-dominated design there exists some q such that the qth performance measure of the selected design is at least d_q better.

On the other hand, if the true performance measures are deviated no more than d_q for some q, then the procedure cannot rank any two systems correctly with high confidence and consequently the selected systems may be dominated (even though for practical purposes they are considered to be indifferent with respect to a particular performance measure). For TSP, under the assumption that the performance measures deviated more than d_q for all q, the P(CS) is the probability of identifying the correct Pareto set.

14.3 Empirical Experiments

In this experiment, we test the multi-objective selection procedures SRSIP and TSP. In order to compare with other known Pareto-set-selection procedures we use a similar setting in [Lee et al. (2004)].

14.3.1 *Experiment 1: The Parameter* $m_p = 2$

The number of systems under consideration $k = 5$ and the number of performance measures $\omega = 3$. The means to generate the systems are listed in Table 14.2. The variance of all systems is 9^2. The indifference amount of performance measures 1, 2, 3, are $d_1 = 1, d_2 = 4$, and $d_3 = 8$. From Table 14.2, system 3 is dominated by systems 1 and 2; system 4 is dominated by system 2; system 5 is dominated by systems 2 and 4; and systems 1 and 2 are non-dominated systems.

With $m_p = 2$, the procedure will select non-dominated systems from two performance measures. We perform 10,000 independent experiments to obtain the actual P(CS). The number of times we successfully selected the true best system in each performance measure is counted among the 10,000 independent experiments. P(CS), the correct selection proportion, is then obtained by dividing this number by 10,000.

Table 14.3 lists the experimental results of SRSIP. Note that the correct selection under performance measures 1, 2, and 3 are systems 1, 2, and 1, respectively. The observed P(CS) under performance measures 1, 2, and 3 are, respectively, 0.6752, 0.9203, and 0.9999; which are, as expected, less than, around, and greater than the nominal value of 0.90. With $m_p = 2$ the allocated sample sizes are neither large enough nor distributed correctly to provide the required probability guarantee to rank systems correctly under performance measure 1, which is likely to have the smallest t score. Consequently, the Pareto set contains only systems 1 and 2 0.8080 fraction of the time. System 3 is incorrectly included in the Pareto set at least 0.0962 fraction of the time. In this case the allocated sample sizes are only large enough to have 90% confidence that the selected system is non-dominated when the system is deviated more than 2 (i.e., 42-40 under performance measure 2). Furthermore, most of the samples are allocated to rank systems 2 and 4. On the other hand, while system 3 deviated from system 1 exactly 2 (i.e., 18-16) under performance measure 1, the samples are not allocated properly to rank systems 1 and 3 with the desired precision.

Table 14.4 shows the sample sizes among the three algorithms: SRSIP,

TOAP, and MOCBA. Under each algorithm, the T column lists the allocated sample size for each system and the θ column lists the proportion of the sample size allocated for each system. The critical constant $h = 2.747$ and with variance $\sigma_{iq}^2 = 9^2$, the theoretical required sample sizes based on performance measure 3 are 10 (i.e., $\lceil (h\sigma_{i1}/d_3)^2 \rceil = \lceil (2.747 * 9/8)^2 \rceil$), 10, 8, 7, and 6, respectively. The theoretical required sample sizes base on performance measure 2 are 39 (i.e., $\lceil (h\sigma_{i2}/d_2)^2 \rceil = \lceil (2.747 * 9/4)^2 \rceil$), 153, 25, 153, and 68, respectively. Hence, the TOAP for systems 1 through 5 are, respectively, 39 (i.e., $\max(10, 39)$), 153, 25, 153, and 68. The average total sample size allocated by the SRSIP procedure is 421, which is slightly less than 436 (i.e., TOAP). This is consistent with the results of [Lee et al. (2004)]. They suggest the reason may be due to the fact that sequential procedures can make use of the sampling information from the previous steps to make decisions regarding the allocation of additional samples. The total sample size of the MOCBA procedure is less than that of SRSIP and TOAP. Note that the goal of the MOCBA procedure is different than SRSIP and TOAP.

14.3.2 *Experiment 2: The Parameter $m_p = 3$*

With $m_p = 3$, the procedure will select non-dominated systems from three performance measures. Tables 14.5 and 14.6 list the results of SRSIP. Tables 14.7 and 14.8 list the results of TSP. For SRSIP, the observed P(CS) under all performance measures are greater than the nominal value of 0.9 with the allocated sample size 1414 less than the theoretical value of 1598. The results indicate that system 1 are non-dominated under two different performance measures. This information may help decision maker to select system 1 since it is non-dominated in more performance measures than other systems. As expected, the observed P(CS) of TSP under all performance measures are greater than those of SRSIP with larger sample sizes. The sample sizes allocated by the TSP procedure are large enough to rank systems with the desired precision when the differences between systems (under all performance measures) are greater than the indifference amounts specified for each performance measure. The Pareto set contains only systems 1 and 2 by SRSIP and TSP, respectively, 0.9622 and 0.9971 fraction of the time. For TSP, systems 3, 4, and 5 are incorrectly included in the Pareto set approximately 0.0021, 0.0002, and 0.0064 fraction of the time, respectively. For both procedures, the allocated sample sizes are close to their theoretical values.

Table 14.2 Means to generate the systems

Systems	Mean_1	Mean_2	Mean_3
1	16	44	56
2	17	40	64
3	18	45	65
4	19	42	66
5	20	43	67

Table 14.3 Proportion of a system having the smallest performance measure of experiment 1

System	PM_1	PM_2	PM_3
1	0.6752	0.0045	0.9999
2	0.2065	0.9203	0.0001
3	0.0962	0.0008	0
4	0.0149	0.0545	0
5	0.0072	0.0199	0

14.3.3 *Experiment 3: The Parameter* $m_p = 3$

In this experiment, not all non-dominated systems have the best performance measure in at least one objective. The means to generate the systems are listed in Table 14.9. The variance of all systems is 9^2. From Table 14.9, systems 4 and 5 are dominated by systems 2 and 3; and systems 1, 2, and 3 are non-dominated systems. Tables 14.10 and 14.11 list the results of TSP. The procedure correctly selected systems 1, 2, and 3 as non-dominated systems 0.9669 fraction of the time; and systems 4 and 5 are incorrectly included in the Pareto set approximately 0.027 and 0.064 fraction of the time, respectively. The TSP is derived based on the least favorable configuration and is conservative. Consequently, it achieves high P(CS) with large sample sizes. For example, the second performance of systems 2 and 3 are 39 and 41, respectively; both values deviated more than the indifference amount (i.e., $d_1 = 1$) from 44 (the second performance measure of system 1). Hence, the allocated sample sizes are sufficient to rank systems 1 and 2, and systems 1 and 3 with the required precision.

14.4 Summary

In this chapter, we present a framework for the ranking and selection problem when the systems are evaluated with more than one performance measure. The procedure incorporates the concept of Pareto optimality into the

Table 14.4 Comparison of SRSIP with TOAP and MOCBA

System	SRSIP		TOAP		MOCBA	
No.	T	θ	T	θ	T	θ
1	82	19.5	39	8.9	88	29.5
2	132	31.3	153	34.9	110	37.1
3	53	12.6	25	5.7	22	7.5
4	96	22.8	153	34.9	48	16.0
5	58	13.8	68	15.6	29	9.9
Total	421	100	436	100	297	100

Table 14.5 Proportion of a system having the smallest performance measure of experiment 2 by SRSIP

System	PM_1	PM_2	PM_3
1	0.9370	0.0013	0.9999
2	0.0426	0.9793	0
3	0.0142	0.0007	0.0001
4	0.0050	0.0135	0
5	0.0012	0.0052	0

Table 14.6 Comparison of SRSIP with TOAP

System	SRSIP		TOAP	
No.	T	θ	T	θ
1	454	32.1	612	38.3
2	532	37.6	612	38.3
3	164	11.6	153	9.6
4	175	12.4	153	9.6
5	89	6.3	68	4.2
Total	1414	100	1598	100

ranking and selection scheme, and attempts to find non-dominated systems in the Pareto set rather than a single "best" system. We present two procedures to solve the problem. The first procedure attempts to select an incomplete Pareto set, i.e., all non-dominated systems that are the best in at least one performance measure; whereas the second procedure attempts to select the Pareto set, i.e., all non-dominated systems. These procedures are versatile and easy to state. However, the TSP procedure is derived based on the least favorable configuration and is conservative.

Table 14.7 Proportion of a system having the smallest performance measure of experiment 2 by TSP

System	PM_1	PM_2	PM_3
1	0.9654	0	1
2	0.0345	0.9994	0
3	0.0001	0	0
4	0	0.0006	0
5	0	0	0

Table 14.8 Comparison of TSP with TOAP

System	TSP		TOAP	
No.	T	θ	T	θ
1	613	20.0	612	20.0
2	610	20.0	612	20.0
3	611	20.0	612	20.0
4	607	20.0	612	20.0
5	612	20.0	612	20.0
Total	3053	100	3060	100

Table 14.9 Means to generate the systems

Systems	$Mean_1$	$Mean_2$	$Mean_3$
1	16	44	56
2	18	39	64
3	17	41	65
4	19	42	66
5	20	43	67

Table 14.10 Proportion of a system having the smallest performance measure of experiment 3

System	PM_1	PM_2	PM_3
1	0.9675	0	1
2	0	0.9997	0
3	0.0325	0.0003	0
4	0	0	0
5	0	0	0

Table 14.11 Comparison of TSP with TOAP

System	TSP		TOAP	
No.	T	θ	T	θ
1	613	20.0	612	20.0
2	610	20.0	612	20.0
3	611	20.0	612	20.0
4	607	20.0	612	20.0
5	612	20.0	612	20.0
Total	3053	100	3060	100

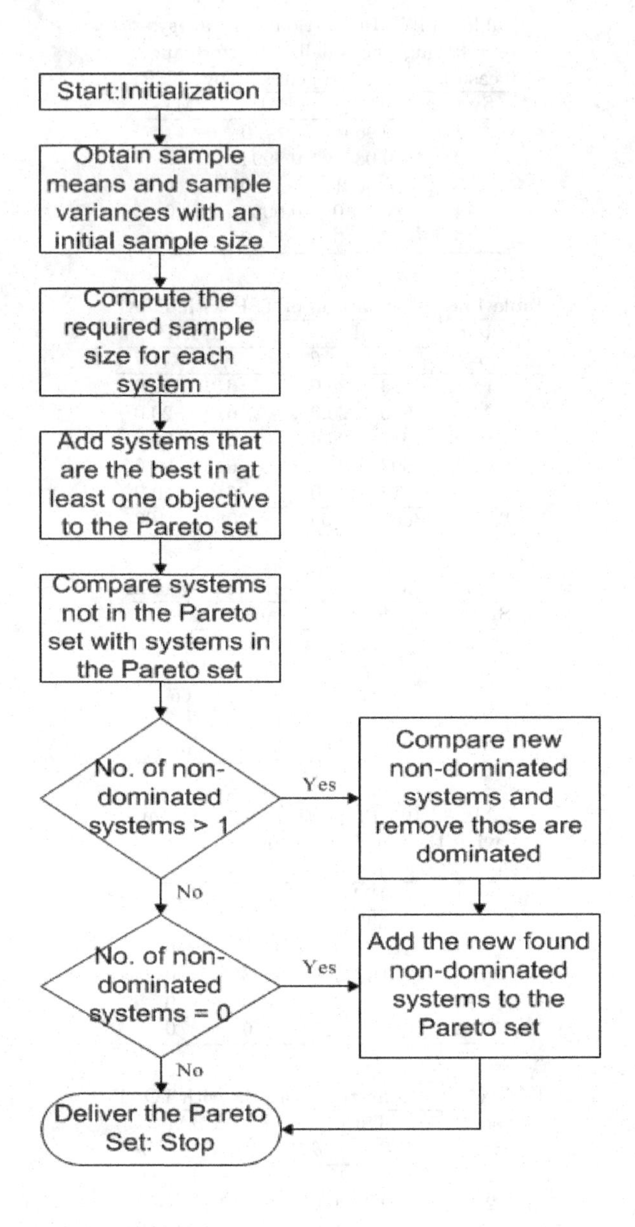

Fig. 14.1 High-level flowchart of TSP

Chapter 15

Generic Selection with Constraints

[Andradóttir and Kim (2010)] develop procedures to find the system with the best primary performance measure in the presence of a stochastic constraint on a secondary performance measure, i.e., selection with constraints. Their fully sequential procedures increase one sample for each system that is still under consideration at each iteration and eliminate systems from further simulation when the partial sum is greater than a threshold. These procedures are efficient in terms of sample sizes but not necessarily in terms of runtime. [Morrice and Butler (2006)] extend the multiple-attribute-utility theorem [Butler et al. (2010)] to perform selection with constraints. However, in some cases, it is not straightforward to assign utility scores.

There are no hard constraints in selecting the Pareto set; however in selection with constraints, systems that do not satisfy the constraints will be removed from further consideration. The bounds of current selection with constraints are based on user specified values of the underlying performance measures. Hence, it is possible that there is no feasible solutions with a given performance bound. We propose using the relative performance measures as the constraints. That is, systems with the performance measure within a user-specified amount of the unknown best are considered as feasible systems. In this setting, there always will be feasible solutions with a given indifference amount of a particular performance measure. We call this a generic selection-with-constraint procedure because the constraints can be a combination of user specified values and/or user specified indifference amounts, which can be either absolute or relative.

15.1 Methodologies

In this section, we formulate a generic selection-with-constraints procedure. Please refer to Section 14.1 for notations. As with most selection procedures, the proposed selection procedures require the input data to be i.i.d. normal. However, the variance can be different across systems. Many performance measures of interest are taken over some average of a sample path or a batch of samples. Thus, many applications tend to have a normally distributed simulation output. If the non-normality of the samples is a concern, users can use batch means to "manufacture" samples that appear to be i.i.d. normal, as determined by the tests of independence and normality (see, e.g., [Chen and Kelton (2007)]). In the selection procedures described below, the sampling operations can be carried out independently across systems. Hence, one can deploy the selection procedures in a parallel and distributed environment.

15.1.1 *Multi-Objective Selection*

When there are multiple selection criteria, we use the *Bonferroni inequality* to compute the required precision in each objective (performance measure). For example, if there are ω objectives and the desired overall P(CS) is P^*, then the required P(CS) in each objective is $P^*_\omega = 1 - (1 - P^*)/\omega$. Let $q = 1$ be the primary performance measure. In selection with constraints, the goal is to find the best feasible system

$$\arg \min_{i=1,\ldots,k} \mu_{i1}$$
$$\text{s.t. } \mu_{iq} < \mu_{0q} + d^*_q \text{ for } q = 2, 3, \ldots, \omega.$$

Here μ_{0q} and d^*_q are, respectively, the user specified standard and indifference amount of performance measure g and each constraint will hold with probability at least P^*_ω.

When there are hard constraints, say C_q for some $q \in \{1, 2, \ldots, \omega\}$, systems i having $\mu_{iq} \geq C_q$ will be removed from consideration. To incorporate the indifference-zone approach, C_q can be written as $\mu_{0q} + d^*_q$, where μ_{0q} is the soft constraint of performance measure q. It is ideal to select desirable systems i having $\mu_{iq} < \mu_{0q}$; however, it is acceptable if $\mu_{0q} \leq \mu_{iq} < \mu_{0q} + d^*_q$. Hence, systems i having $\mu_{iq} < \mu_{0q} + d^*_q$ are feasible systems. With the indifference-zone approach, desirable systems will be included in the subset with high probability; the acceptable systems may be included in the subset, but there is no probability guarantee. Let $\bar{X}_{iq}$

denote the sample mean of the qth performance measure of system i. The proposed procedure will treat systems i having $\bar{X}_{iq} < \mu_{0q} + d_q^*/2$ as feasible systems.

With $C_q = \mu_{0q} + d_q^*$ as the constraints, users need to specify the indifference amount d_q^* and the value μ_{0q}. However, users may not have the priori information to specify a meaningful μ_{0q}. We propose new constraints using the technique of comparison with the best. With the indifference-zone approach, we intend to have $C_q = \mu_{i_1q} + d_q^*$. It is ideal to select desirable systems i having $\mu_{iq} = \mu_{i_1q}$; however, it is acceptable if $\mu_{i_1q} < \mu_{iq} < \mu_{i_1q} + d_q^*$. Hence, systems i having $\mu_{iq} < \mu_{i_1q} + d_q^*$ are feasible systems. However, the best system (i_1) and its mean (μ_{i_1q}) are unknown. Let system b_l be the system having the lth smallest sample mean of performance measure q, i.e., $\bar{X}_{b_1q} \leq \bar{X}_{b_2q} \leq \cdots \leq \bar{X}_{b_kq}$. The proposed procedure will treat systems b_l having $\bar{X}_{b_lq} < \bar{X}_{b_1q} + d_q^*/2$ as feasible systems. That is, the bounds $\mu_{0q} + d_q^*/2$ is replaced by $\bar{X}_{b_1q} + d_q^*/2$. This is analogous to Comparison With the Standard vs. Comparison With a Control. The true means of the control μ_{i_1q} are unknown and need to be estimated. Using comparison with a control as constraint also allows us to exploit using common random number to increase efficiency.

15.1.2 *A Generic Selection-With-Constraints Procedure*

In this section, we define additional notations and formulate a selection-with-constraints problem. We call this the SWCG (Selection With Constraints - Generic) Procedure. Let $0 \leq \omega' \leq \omega$, performance measures $q = 2, 3, \ldots, \omega'$ will be subjected to the constraints of comparison with a control (the best being the control) and performance measures $q = \omega' + 1, \omega' + 2, \ldots, \omega$ will be subjected to the constraints of comparison with the standard. That is, the goal is to find the best feasible system

$$
\begin{aligned}
\arg \min_{i=1,\ldots,k} \ & \mu_{i1} \\
\text{s.t.} \ & \mu_{iq} < \mu_{i_1q} + d_q^* \text{ for } q = 2, 3, \ldots, \omega' \\
& \mu_{iq} < \mu_{0q} + d_q^* \text{ for } q = \omega' + 1, \omega' + 2, \ldots, \omega.
\end{aligned}
$$

Let θ be the set of systems that are still under consideration and is initialized to include all k systems. Note that θ will be changed from iteration to iteration. Let $\bar{X}_{iq}$ denote the sample mean of the qth performance measure of system i. Let $\bar{X}_{b_1q} = \min_{i\in\theta} \bar{X}_{iq}$ and let $U(\bar{X}_{b_1q})$ be the upper

P^* confidence limits of μ_{b_1q}. Then

$$d_{iq} = \max(d_q^*, \bar{X}_{iq} - U(\bar{X}_{b_1q})) \text{ for } q = 1, 2, \ldots, \omega'. \quad (15.1)$$

For the primary performance measure, the sample size will be computed according to Eq. (4.2) with d^* replaced by d_{i1}. For the secondary performance measures, we are performing a selecting only and/or all the best systems (for $q = 2, 3, \ldots, \omega'$) or Comparison With the Standard (for $q = \omega' + 1, \omega' + 2, \ldots, \omega$). Furthermore, to take into account the difference of sample means when computing the required sample sizes for $q = \omega' + 1, \omega' + 2, \ldots, \omega$, we use the control distance

$$d_{iq} = \begin{cases} \max(d_q^*, L(\bar{X}_{iq}) - \mu_{0q}) & \text{when } \bar{X}_{iq} > \mu_{0q} \\ \max(d_q^*, \mu_{0q} - U(\bar{X}_{iq})) & \text{when } \bar{X}_{iq} \le \mu_{0q}. \end{cases} \quad (15.2)$$

Here $L(\bar{X}_{iq})$ and $U(\bar{X}_{iq})$ are, respectively, the lower and upper P^* confidence limits of μ_{iq}.

There is no more than $\beta = (1 - P^*)/\omega$ probability that a system that is infeasible with respect to performance measure q be declared feasible. Consequently, with these sample sizes, we have (approximately) P^* confidence that the selected system is a feasible d_q^*-near-best system.

To improve efficiency of the procedure, we also compare systems that are still under consideration with systems that are found to be feasible. Note that θ contains systems that are declared to be feasible and systems that need more sampling to verify whether they are feasible. Systems that are found to be infeasible are excluded from θ. Let $S_{iq}^2(N_i)$ denote the sample variance of the qth performance measure of system i with sample size N_i. When X_{iqj} is i.i.d. normal, it is known that the random variable $Y_{i1j} = \bar{X}_{i1} - \bar{X}_{j1}$ $(i \neq j)$ has approximately a t distribution with f_{ij} d.f., where

$$f_{ij} = \frac{(S_{i1}^2(N_i)/N_i + S_{j1}^2(N_j)/N_j)^2}{(S_{i1}^2(N_i)/N_i)^2/(N_i - 1) + (S_{j1}^2(N_j)/N_j)^2/(N_j - 1)}; \quad (15.3)$$

see [Law (2014)] for details. Since f_{ij} will not, in general, be an integer, interpolation will probably be necessary. The procedure will eliminate system j such that $\bar{X}_{j1} > \bar{X}_{i1} + w_{ij}$ for some feasible system i. Here $w_{ij} = t_{f_{ij},P}\sqrt{S_{i1}^2(N_i)/N_i + S_{j1}^2(N_j)/N_j}$ is the one-tailed $P(= (P_\omega^*)^{1/(k-1)})$ confidence interval half width. The SWCG procedure is as follows.

A Generic Selection-With-Constraints Procedure

(1) Let $N_{i,r}$ be the sample size allocated for system i and $\bar{X}_{iq,r}$ be the sample mean of the qth performance measure of system i at the rth iteration. Simulate n_0 samples for all systems. Set the iteration number $r = 0$, and $N_{1,r} = N_{2,r} = \cdots = N_{k,r} = n_0$. Specify the value of the indifference amount d_q for $q = 1, 2, \ldots, \omega$, the soft constraints μ_{0q} for $q = \omega' + 1, \omega' + 2, \ldots, \omega$, and the required precision P^*. Let θ be the set of systems that are still under consideration and is initialized to include all k systems. Compute $P^*_\omega = 1 - (1 - P^*)/\omega$ and $P = (P^*_\omega)^{1/(k-1)}$. The critical constants h_1 and h_3 are obtained with k, n_0, and P^*_ω.

(2) Calculate the sample means and sample variances for each performance measure of each system. Obtain the index $b_1 (= \arg\min_{i \in \theta} \bar{X}_{i1})$, i.e., the system having the smallest sample mean of the primary performance measure.

(3) Calculate the new sample sizes

$$N_{i1,r+1} = \max(n_0, \lceil (h_1 S_{i1}(N_{i,r})/d_{i1})^2 \rceil), \text{ for } i \in \theta,$$

$$N_{iq,r+1} = \max(n_0, \lceil (2 h_3 S_{iq}(N_{i,r})/d_{iq})^2 \rceil),$$

$$\text{for } i \in \theta \text{ and } q = 2, 3, \ldots, \omega',$$

and

$$N_{iq,r+1} = \max(n_0, \lceil (2 t_{N_{i,r}-1,P} S_{iq}(N_{i,r})/d_{iq})^2 \rceil),$$

$$\text{for } i \in \theta \text{ and } q = \omega' + 1, \omega' + 2, \ldots, \omega.$$

Here d_{iq}, for $i \in \theta$ and $q = 1, 2, \ldots, \omega'$, are computed according to Eq. (15.1), d_{iq}, for $i \in \theta$ and $q = \omega' + 1, \omega' + 2, \ldots, \omega$, are computed according to Eq. (15.2), and $S^2_{iq}(N_{i,r})$ is the sample variance of the qth performance measure of system i with sample size $N_{i,r}$.

(4) Let $w_{ib_1q} = t_{f_{ib_1},P} \sqrt{S^2_{iq}(N_{i,r})/N_{i,r} + S^2_{b_1q}(N_{b_1,r})/N_{b_1,r}}$. For each system $i \in \theta$, if $\bar{X}_{iq} > \bar{X}_{b_1q} + w_{ib_1q}$ for some $q = 2, 3, \ldots, \omega$, then remove system i from θ.

(5) Let $w_{i0q} = t_{N_{i,r}-1,P} S_{iq}(N_{i,r})/\sqrt{N_{i,r}}$. For each system $i \in \theta$, if $\bar{X}_{iq} > \mu_{0q} + w_{i0q}$ for some $q = \omega' + 1, \omega' + 2, \ldots, M$, then remove system i from θ. For each system $i \in \theta$, if $\bar{X}_{iq} < \mu_{0q} - w_{i0q}$ for some $q = \omega' + 1, \omega' + 2 \ldots, \omega$, then set $N_{iq,r+1} = N_{iq,r}$.

(6) For each $i \in \theta$ having $N_{iq,r+1} = N_{iq,r}$, for $q = 1, 2, \ldots, \omega$, then for each $j \in \theta$ compute $w_{ijq} = t_{f_{ij},P}\sqrt{S^2_{iq}(1, N_{i,r})/N_{i,r} + S^2_{jq}(1, N_{j,r})/N_{j,r}}$. If $\bar{X}_{j1} > \bar{X}_{i1} + w_{ij1}$, then remove system j from θ.
(7) If $\theta = \emptyset$, then terminate the program. There is no feasible solution.
(8) Let

$$N_{i,r+1} = \max_{q=1,\ldots,\omega} N_{iq,r+1}$$

and let

$$N'_{i,r+1} = \min_{q=1,\ldots,\omega} \{N_{iq,r+1} - N_{iq,r} | N_{iq,r+1} - N_{iq,r} > 0\}.$$

If $N_{i,r+1} \le N_{i,r}$, for $i = 1, 2, \ldots, k$, go to step 10.
(9) Simulate additional $\min(N'_{i,r+1}, \lceil (N_{i,r+1} - N_{i,r})^+/2 \rceil)$ samples for systems $i \in \theta$. Set $r = r + 1$. Go to step 2.
(10) Select system $b_1 = \min_{i \in \theta} \bar{X}_{i1}$.

In the case that there is only one secondary performance measure, a tighter lower bound can be achieved. Let β_1 be the probability that an unacceptable system is included in the subset and let β_2 be the probability that the best feasible system is not selected when compared to other feasible systems in isolation. When $\beta_1 = \beta_2 = \beta$, [Andradóttir and Kim (2010)] show that

$$\mathrm{P(CS)} \ge 2(1-\beta)^{(k-1)/2} - 1 - \beta = P^*. \tag{15.4}$$

15.1.3 *Variance as the Constraint*

[Batur and Choobineh (2010)] discuss a selection procedure based on both the mean and variance. In this section, we implement a procedure with variance as the constraint. The goal is to find the best feasible system

$$\arg \min_{i=1,\ldots,k} \mu_i$$
$$\text{s.t. } \sigma_i^2/\sigma_{j_1}^2 < d_r^* \text{ for } i = 1, 2, 3, \ldots, k.$$

Again, for the primary performance measure, the sample size will be computed according to Eq. (4.2) with d^* replaced by $d_i = \max(d^*, \bar{X}_i - U(\bar{X}_{b_1}))$ for $i = 1, 2, 3, \ldots, k$. For the secondary performance measure, we are performing a selecting only and/or all the best systems with respect to variance. From Section 10.2.3, once the indifference ratio d_r^* is specified, the required sample size $N_{d_r^*}$ (i.e. d.f.) to select feasible systems can be determined. Hence, we can set the initial sample sizes $n_0 = N_{d_r^*}$ and then the procedure is simplified to select the best system with respect to the primary performance measure. The procedure is as follows.

SWCG with variance as the constraint

(1) Let θ be the set of systems that are still under consideration and is initialized to include all k systems. Specify the value of the indifference amounts d^* and d_r^* for the primary and secondary performance measures and the required precision P^*. Compute $P_2^* = 1 - (1 - P^*)/2$. Determine the required sample size $N_{d_r^*}$. Set $n_0 = N_{d_r^*}$. The critical constant h_1 is obtained with k, n_0, and P_2^*.
(2) Let $N_{i,r}$ be the sample size allocated for system i at the rth iteration. Simulate n_0 samples for all systems. Set the iteration number $r = 0$, and $N_{1,r} = N_{2,r} = \cdots = N_{k,r} = n_0$.
(3) Calculate the sample variance of each system. Obtain the index $j_1(= \arg\min_{i\in\theta} S_i^2)$, i.e., the system having the smallest variance. Remove systems i such that $(S_i/S_{j_1})^2 > \sqrt{d_r^*}$ from θ.
(4) Calculate the sample mean and sample variance of each system. Obtain the index $b_1(= \arg\min_{i\in\theta} \bar{X}_i)$, i.e., the system having the smallest sample mean.
(5) Calculate the new sample sizes

$$N_{i1,r+1} = \max(n_0, \lceil (h_1 S_i(N_{i,r})/d_{i1})^2 \rceil), \text{ for } i \in \theta.$$

(6) If $N_{i,r+1} \leq N_{i,r}$, for $i = 1, 2, \ldots, k$, go to step 8.
(7) Simulate additional $\lceil (N_{i,r+1} - N_{i,r})^+ \rceil$ samples for systems $i \in \theta$. Set $r = r + 1$. Go to step 4.
(8) Select system $b_1 = \min_{i\in\theta} \bar{X}_i$.

15.1.4 *Variance as the Primary Performance Measure*

In this section, we implement a procedure with variance as the primary performance measurement and mean as the constraint. The goal is to find the best feasible system

$$\arg \min_{i=1,\ldots,k} \sigma_i^2$$
$$\text{s.t. } \mu_i < \mu_{i_1} + d^* \text{ for } i = 1, 2, 3, \ldots, k.$$

We first obtain the required sample size $N_{d_r^*}$ with the given d_r^*, k, and P_2^*. We then set the initial sample sizes $n_0 = N_{d_r^*}$ and consequently the procedure is simplified to determine the feasible systems. The procedure is as follows.

SWCG with Variance as Primary Performance Measure

(1) Let θ be the set of systems that are still under consideration and is initialized to include all k systems. Specify the value of the indifference amounts d_r^* and d^* for the primary and secondary performance measures and the required precision P^*. Compute $P_2^* = 1-(1-P^*)/2$ and $P = (P_2^*)^{1/k}$. Determine the required sample size $N_{d_r^*}$. Set $n_0 = N_{d_r^*}$. The critical constant h_3 is obtained with k, n_0, and P_2^*.
(2) Let $N_{i,r}$ be the sample size allocated for system i at the rth iteration. Simulate n_0 samples for all systems. Set the iteration number $r = 0$, and $N_{1,r} = N_{2,r} = \cdots = N_{k,r} = n_0$.
(3) Calculate the sample means and sample variances. Obtain the index $b_1 (= \arg\min_{i\in\theta} \bar{X}_i)$, i.e.,the system having the smallest sample means.
(4) Calculate the new sample sizes $N_{i,r+1} = \max(n_0, \lceil (2h_3 S_i(N_{i,r})/d_{i1})^2 \rceil)$, for $i \in \theta$.
(5) If $N_{i,r+1} \leq N_{i,r}$, for $i = 1, 2, \ldots, k$, go to step 7.
(6) Simulate additional $\lceil (N_{i,r+1} - N_{i,r})^+ \rceil$ samples for systems $i \in \theta$. Set $r = r + 1$. Go to step 3.
(7) For each $i \in \theta$, remove i from θ when $\bar{X}_i > \bar{X}_{b_1} + d^*/2$. Select system $b_1 = \min_{i\in\theta} S_i^2(N_{i,r})$.

15.2 Empirical Experiments

In this section, we present some empirical results of performing selection with constraints.

15.2.1 *Selection With Constraints*

In this experiment, we test the Selection-With-Constraints Procedure. We use a similar setting in [Andradóttir and Kim (2010)]. The number of systems under consideration $k = 25$. There are two performance measures, i.e., $\omega = 2$. The primary performance measure

$$\mu_{i1} = \begin{cases} 0, & i = 1, 2, \ldots, b-1, \\ -d^*, & i = b, \\ 0, & i = b+1, \ldots, b+a, \\ (1-i)d^*, & i = b+a+1, \ldots, k; \end{cases}$$

and the secondary performance measure

$$\mu_{i2} = \begin{cases} -\epsilon, & i = 1, 2, \ldots, b, \\ 0, & i = b+1, \ldots, b+a, \\ \epsilon, & i = b+a+1, \ldots, k. \end{cases}$$

The primary and secondary performance measures of each system are independent. The variances of both performance measures of each system are 1. The initial sample size n_0 is set to $n_0 = 20$. Moreover, the indifference amount of the primary performance measure d_1^* and the tolerance ϵ are set to $1/\sqrt{n_0}$. Furthermore, the soft constraint $\mu_{02} = -\epsilon$ and the indifference amount of the secondary performance measure $d_2^* = 2\epsilon$, thus, the hard constraint $C_2 = \epsilon$. The desired P(CS) is set to $P^* = 0.95$. We make 10,000 independent replications to obtain the observed P(CS). Note that with $\omega = 2$ performance measures $P_\omega^* = 1-(1-P^*)/2 = 0.975$. The critical constants $h_1 = 4.60$ of Eq. (4.2) and $h_3 = 6.23$ of Eq. (4.4) are obtained with $n_0 = 20$, $k = 25$, and $P_\omega^* = 0.975$. Furthermore, $P = P_\omega^{*1/(k-1)} \approx 0.999$.

In cases that there is only one secondary performance measure, a tighter lower bound can be achieved. Let β_1 be the probability that an unacceptable system is included in the subset and let β_2 be the probability that the best feasible system is not selected when compared to other feasible systems in isolation. If $\beta_1 = \beta_2 = \beta$, Eq. (15.4) can be used. When $P^* = 0.95$, $\beta \approx 0.002$. Hence, for the SWCGO (SWCG with the parameters optimized) procedure, we set $P = (1 - 0.002) = 0.998$. Furthermore, the critical constants $h_1 = 4.22$ of Eq. (4.2) and $h_3 = 5.88$ of Eq. (4.4) are obtained with $k = 25$ and $P_\omega^* = 0.954$ (i.e. 0.998^{24}).

Tables 15.1 and 15.2, respectively, show the experimental results when the constraint is the standard and the unknown best. We list the observed P(CS) and the average of the allocated sample sizes T, the standard error of T and the average number of iterations of *SWCG*, *SWCGO*, and AK+ [Andradóttir and Kim (2010)]) for various numbers of acceptable systems a when $k = 25$ and $a+b = 13$. The observed P(CS)'s are all greater than the specified nominal value. However, the allocated sample sizes with unknown $\mu_{i_1 2}$ as the constraint are much larger than those with known constrain μ_{02}. Both SWCG and SWCGO are conservative, achieve high precision with large sample sizes. AK+ is efficient in terms of sample sizes, however, it requires many more iterations than SWCG and SWCGO.

Table 15.1 The performance of selection with constraints: given the standard and indifference amount

		$b=13$ $a=0$	$b=12$ $a=1$	$b=10$ $a=3$	$b=7$ $a=6$	$b=3$ $a=10$	$b=1$ $a=12$
SWCG	*PCS*	0.982	0.979	0.978	0.986	0.982	0.987
	T	6494	6335	6065	5646	5096	4789
	std(T)	602	623	593	511	454	440
	Iter	13	13	13	12	11	9
SWCGO	*PCS*	0.976	0.972	0.971	0.990	0.984	0.991
	T	5816	5711	5483	5197	4774	4585
	std(T)	489	468	458	408	387	377
	Iter	13	13	13	12	11	10
AK+	*PCS*	0.960	0.962	0.963	0.963	0.966	0.968
	T	3976	3749	3726	3686	3615	3581
	$\hat{Iter}$	140	130	130	128	125	124

Table 15.2 The performance of selection with constraints: given the indifference amount only

		$b=13$ $a=0$	$b=12$ $a=1$	$b=10$ $a=3$	$b=7$ $a=6$	$b=3$ $a=10$	$b=1$ $a=12$
SWCG	*PCS*	0.967	0.966	0.980	0.979	0.996	0.996
	T	11877	11512	10810	9929	9214	9847
	std(T)	902	954	993	1108	1377	1806
	Iter	11	11	11	11	10	6
SWCGO	*PCS*	0.963	0.959	0.969	0.987	0.990	1.000
	T	10586	10290	9712	8998	8413	8999
	std(T)	869	907	930	1044	1200	1630
	Iter	11	11	11	11	10	7

15.2.2 *Variance as the Constraint*

In this experiment, we test the Selection-With-Constraints procedure using variance as the constraint. We use a similar setting in [Batur and Choobineh (2010)]. However, the goal of the procedure is different than that of [Batur and Choobineh (2010)], which is aim to obtain a Pareto set, i.e., the set of non-dominated systems. The number of systems under consideration $k = 5$. The experiment configuration are listed in Table 15.3. The CSM and CSV columns list, respectively, the best system when the primary performance measure is the mean and variance. For example, system 5 is the best system when mean is the primary performance measure under configuration 4 because all other systems do not satisfy the variance constraint. The required P(CS) $P^* = 0.95$, the absolute indifference amount $d^* = 0.1$ and the relative indifference amount $d_r^* = 1.21$.

With $\delta_r = \sqrt{d_r^*} = 1.1, P_\omega^* = 1 - (1 - P^*)/2 = 0.975$, and $k = 5$, the required sample sizes are $n_e = 3881$ and $n_f = 2636$. Hence, we set $n_0 = \max(n_e, n_f) = 3881$. Furthermore, with $n_0 = 3881, P_\omega^* = 0.975$, and $k = 5$, the critical constant $h_1 = 3.456$. Note that the theoretical sample size to achieve P(CS) of P_ω^* of the best mean is $\lceil (h_1 S/d^*)^2 \rceil = \lceil (3.456 * 1.1/0.1)^2 \rceil = 1446 < n_0 = 3881$. Consequently, the initial sample size n_0 (to achieve P(CS) of P_ω^* of the variance constraint) is already large enough to achieve P^*.

We then solve Eq. (15.4) with $k = 5$ and $P^* = 0.95$ to obtain $\beta \approx 0.01$. Hence, for the SWCGO procedure, we set $P = (1 - 0.01) = 0.99$. With $\delta_r = \sqrt{d_r^*} = 1.1, P_\omega^* = 0.961$ (i.e. 0.99^4), and $k = 5$, the required sample sizes are $n_e = 3495$ and $n_f = 2270$. Hence, we set $n_0 = \max(n_e, n_f) = 3495$. Furthermore, with $n_0 = 3495, P_2^* = 0.961$, and $k = 5$, the critical constant $h_1 = 3.206$. The theoretical sample size to achieve P(CS) of P_ω^* of the best mean is $\lceil (h_1 S/d^*)^2 \rceil = \lceil (3.206 * 1.1/0.1)^2 \rceil = 1244 < n_0 = 3495$. Again, the initial sample size n_0 is already large enough to achieve P^*.

Table 15.4 shows the results. The observed P(CS)'s are all greater than the specified nominal value. The required sample size to achieve P(CS) of the variance constraint is based on the LFC and is conservative.

15.2.3 *Variance as the Primary Performance Measure*

In this experiment, we test the Selection-With-Constraints procedure using mean as the constraint and variance as the primary performance measure. The configurations of the systems are the same as the previous experiment. In this configuration, we set $n_0 = n_f = 2636$. Furthermore, with $n_0 = 2636, P = 0.975$, and $k = 5$, the critical constant $h_3 = 4.281$. The theoretical sample size to achieve P(CS) of P_ω^* of the best mean is $\lceil (h_3 S/(d^*/2))^2 \rceil = \lceil (4.281 * 1.1/(0.1/2))^2 \rceil = 8871$, which is greater than $n_0 = 2636$. Consequently, the initial sample size n_0 is not large enough to achieve P^*.

For the SWCGO procedure, we set $n_0 = n_f = 2270$. With $n_0 = 2270, P_\omega^* = 0.961$, and $k = 5$, the critical constant $h_3 = 4.067$. The theoretical sample size to achieve P(CS) of P_ω^* of the best mean is $\lceil (h_3 S/(d^*/2))^2 \rceil = \lceil (4.067 * 1.1/(0.1/2))^2 \rceil = 8006$, which is greater than $n_0 = 2270$. Hence, the initial sample size n_0 is not large enough to achieve P^*.

Table 15.5 shows the results. The observed P(CS)'s are all greater

Table 15.3 Configurations of the systems tested

		Sys. 1	Sys. 2	$\cdots$	Sys. i	$\cdots$	Sys. k	CSM	CSV
Config 1	Mean	1	$1+d^*$	$\cdots$	$1+d^*$	$\cdots$	$1+d^*$	1	1
	Variance	1	d_r^*	$\cdots$	d_r^*	$\cdots$	d_r^*		
Config 2	Mean	1	$1+d^*$	$\cdots$	$1+d^*$	$\cdots$	$1+d^*$	1	1
	Variance	1	1	$\cdots$	1	$\cdots$	1		
Config 3	Mean	1	1	$\cdots$	1	$\cdots$	1	1	1
	Variance	1	d_r^*	$\cdots$	d_r^*	$\cdots$	d_r^*		
Config 4	Mean	1	$1+d^*$	$\cdots$	$1+(i-1)d^*$	$\cdots$	$1+(k-1)d^*$	5	1
	Variance	$d_r^{*(k-1)}$	$d_r^{*(k-2)}$	$\cdots$	$d_r^{*(k-i)}$	$\cdots$	1		
Config 5	Mean	1	$1+d^*$	$\cdots$	$1+(i-1)d^*$	$\cdots$	$1+(k-1)d^*$	1	1
	Variance	1	d_r^*	$\cdots$	$d_r^{*(i-1)}$	$\cdots$	$d_r^{*(k-1)}$		

Table 15.4 The performance of selection with constraints (with variance as the constraint)

Config		1	2	3	4	5
SWCG	PCS	1.000	0.9952	0.9978	0.9979	1.000
	T	19405	19405	19405	19405	19405
	$std(T)$	0	0	0	0	0
	$Iter$	1	1	1	1	1
SWCGO	PCS	1.000	0.9926	0.9963	0.9984	1.000
	T	17475	17475	17475	17475	17475
	$std(T)$	0	0	0	0	0
	$Iter$	1	1	1	1	1

Table 15.5 The performance of selection with constraints (with mean as the constraint)

Config		1	2	3	4	5
SWCG	PCS	1.000	0.9982	0.9959	0.9994	1.000
	T	31878	27954	43217	34272	21403
	$std(T)$	7222	5802	390	3915	2573
	$Iter$	3	3	4	3	2
SWCGO	PCS	1.000	0.9962	0.9927	0.9983	1.000
	T	29665	25818	39040	30946	19196
	$std(T)$	6558	5350	376	2843	2406
	$Iter$	3	3	4	2	2

than the specified nominal value. Even though the required sample size to achieve P(CS) of P^* with mean as the constraint is not based on the LFC, the procedure is conservative. First, the SWSG procedure is derived based on the Bonferroni inequality. Second, for each system the same sample size is used to calculate all performance measures and the largest sample size to obtain the required precision P^*_ω of a particular performance measure is used, hence, the obtained precision for all other performance measures are likely to be greater than the nominal value. One approach is to distribute the probability of $1 - P^*$ unequally among ω performance measures to minimize the required sample size such that $\sum(1-P^*_{\omega_q}) = 1-P^*$, where $P^*_{\omega_q}$ is the specified P(CS) for performance measure q. For example, instead of setting $P^*_\omega = 0.975$ when $P^* = 0.95$, we can set $P^*_{\omega_1} = 0.97$ and $P^*_{\omega_2} = 0.98$ when the first performance measure requires more samples to achieve the same level of precision.

15.3 Summary

We have presented a generic selection-with-constraints procedure. The constraints can be a known bound or an unknown bound based on the unknown best. The procedure is developed based on the Bonferroni inequality and is conservative, achieves high P(CS) with large sample sizes. Nevertheless, the procedure is versatile, easy to understand, and simple to implement. Furthermore, inequality Eq. (15.4) developed by [Andradóttir and Kim (2010)] can be used to increase efficiency of the procedure when there is only one secondary performance measure. A generalized inequality when there are more than one secondary performance measures is being developed.

Appendix A

Tables of Critical Constants

Table A.1 Values of critical constant h for the subset selection procedure

				$P^* = 0.90$				$P^* = 0.95$			
k	m	v	c	15	20	25	30	15	20	25	30
8	3	1	1	1.853	1.830	1.817	1.808	2.305	2.268	2.247	2.234
8	3	2	1	0.889	0.882	0.878	0.875	1.243	1.232	1.225	1.221
			2	2.630	2.586	2.561	2.545	3.070	3.007	2.972	2.949
8	3	3	1	0.329	0.326	0.324	0.323	0.657	0.652	0.648	0.646
			2	1.686	1.667	1.656	1.649	2.037	2.011	1.996	1.987
			3	3.621	3.533	3.484	3.453	4.122	4.004	3.939	3.897
9	3	1	1	1.968	1.943	1.929	1.920	2.417	2.380	2.358	2.344
9	3	2	1	1.027	1.019	1.013	1.010	1.375	1.362	1.355	1.350
			2	2.740	2.694	2.668	2.651	3.180	3.115	3.078	3.055
9	3	3	1	0.508	0.504	0.501	0.500	0.828	0.821	0.816	0.814
			2	1.818	1.796	1.784	1.776	2.163	2.133	2.116	2.106
			3	3.723	3.630	3.579	3.546	4.221	4.096	4.028	3.984
10	4	1	1	1.740	1.719	1.707	1.699	2.181	2.146	2.126	2.114
10	4	2	1	0.796	0.790	0.787	0.784	1.135	1.126	1.120	1.117
			2	2.388	2.350	2.329	2.315	2.810	2.754	2.723	2.703
10	4	3	1	0.270	0.268	0.267	0.266	0.578	0.574	0.571	0.570
			2	1.453	1.439	1.431	1.426	1.770	1.751	1.741	1.733
			3	2.977	2.920	2.889	2.869	3.3403	3.325	3.282	3.255
10	4	4	1	0	0	0	0	0.150	0.149	0.148	0.148
			2	0.929	0.922	0.917	0.914	1.223	1.212	1.206	1.202
			3	2.064	2.039	2.024	2.015	2.396	2.362	2.343	2.331
			4	3.888	3.788	3.732	3.696	4.381	4.249	4.175	4.129
10	5	1	1	1.453	1.434	1.423	1.417	1.894	1.862	1.844	1.832
10	5	2	1	0.479	0.475	0.474	0.472	0.815	0.808	0.804	0.802
			2	2.041	2.008	1.989	1.977	2.459	2.408	2.379	2.361
10	5	3	1	0	0	0	0	0.221	0.219	0.218	0.218
			2	1.076	1.067	1.062	1.058	1.392	1.378	1.370	1.365
			3	2.506	2.460	2.434	2.418	2.921	2.855	2.818	2.794
10	5	4	1	0	0	0	0	0	0	0	0
			2	0.510	0.506	0.504	0.503	0.797	0.791	0.788	0.785
			3	1.560	1.544	1.535	1.529	1.871	1.849	1.837	1.829
			4	3.027	2.966	2.932	2.910	3.448	3.364	3.318	3.289

Table A.2 Values of P(CS) of ratio subset selection with $n_0 = 20$

				d^*					
k	m	v	c	1.2	1.4	1.6	1.8	2.0	2.2
8	3	1	1	0.5698	0.7289	0.8399	0.9102	0.9510	0.9740
8	3	2	1	0.9134	0.9626	0.9849	0.9943	0.9978	0.9991
			2	0.2515	0.4283	0.5955	0.7303	0.8275	0.8929
8	3	3	1	0.9376	0.9804	0.9941	0.9983	0.9995	0.9998
			2	0.5221	0.7201	0.8501	0.9243	0.9630	0.9822
			3	0.0656	0.1558	0.2773	0.4096	0.5346	0.6430
9	3	1	1	0.5250	0.6897	0.8107	0.8901	0.9389	0.9667
9	3	2	1	0.7844	0.9016	0.9587	0.9837	0.9938	0.9976
			2	0.2105	0.3772	0.5451	0.6877	0.7950	0.8705
9	3	3	1	0.9088	0.9693	0.9903	0.9970	0.9991	0.9997
			2	0.4518	0.6610	0.8102	0.9007	0.9498	0.9752
			3	0.0491	0.1245	0.2347	0.3622	0.4876	0.6012
10	4	1	1	0.6014	0.7593	0.8651	0.9286	0.9636	0.9822
10	4	2	1	0.8497	0.9416	0.9792	0.9931	0.9977	0.9992
			2	0.3052	0.5018	0.6748	0.8026	0.8862	0.9368
10	4	3	1	0.9481	0.9860	0.9965	0.9991	0.9997	0.9999
			2	0.5944	0.7920	0.9051	0.9600	0.9839	0.9937
			3	0.1180	0.2612	0.4318	0.5937	0.7253	0.8221
10	4	4	1	0.9841	0.9968	0.9994	0.9998	0.9999	0.9999
			2	0.7943	0.9220	0.9733	0.9914	0.9972	0.9991
			3	0.3131	0.5372	0.7225	0.8466	0.9195	0.9591
			4	0.0265	0.0815	0.1744	0.2931	0.4197	0.5394
10	5	1	1	0.7003	0.8378	0.9187	0.9616	0.9825	0.9924
10	5	2	1	0.9182	0.9739	0.9925	0.9979	0.9994	0.9998
			2	0.4401	0.6469	0.7989	0.8934	0.9464	0.9740
10	5	3	1	0.9806	0.9960	0.9992	0.9998	0.9999	0.9999
			2	0.7553	0.9006	0.9642	0.9881	0.9962	0.9988
			3	0.2370	0.4389	0.6299	0.7755	0.8711	0.9294
10	5	4	1	0.9963	0.9994	0.9999	0.9999	1.0000	1.0000
			2	0.9136	0.9763	0.9941	0.9986	0.9996	0.9999
			3	0.5328	0.7553	0.8874	0.9526	0.9812	0.9926
			4	0.0986	0.2371	0.4108	0.5786	0.7157	0.8171

Bibliography

Abramowiz, M. and Stegun, I. A. (1964). *Handbook of Mathematical Functions.* National Bureau of Standards.

Alexopoulos, C., Goldsman, D., Mokashi, A., Nie, R., Sun, Q., Tien, K. W. and Wilson, J. R. (2014). Sequest: A Sequential Procedure for Estimating Steady-State Quantiles. *Proceedings of the 2014 Winter Simulation Conference*, pp. 662–673.

Andradóttir, S. and Kim, S. H. (2010). Fully Sequential Procedures for Comparing Constrained Systems via Simulation. *Naval Research Logistics* **57**, 5, pp. 403–421.

Ankenman, B., Nelson, B. L. and Staum, J. (2008). Stochastic Kriging for Simulation Metamodeling, in *Proceedings of the 2008 Winter Simulation Conference*, pp. 362–370.

Arellano-Valle, R. B. and Genton, M. G. (2007). On the Exact Distribution of Linear Combinations of Order Statistics from Dependent Random Variables. *Journal of Multivariate Analysis* **98**, pp. 1876–1894.

Arellano-Valle, R. B. and Genton, M. G. (2008). On the Exact Distribution of the Maximum of Absolutely Continuous Dependent Random Variables. *Statistics & Probability Letters* **78**, pp. 27–35.

Batur, D. and Choobineh, F. F. (2010). Mean-Variance Based Randking and Selection, in *Proceedings of the 2010 Winter Simulation Conference*, pp. 1160–1166.

Bechhofer, R. E. (1954). A Single-Sample Multiple Decision Procedure for Ranking Means of Normal Populations with Known Variances. *Ann. Math. Statist* **25**, pp. 16–39.

Billingsley, P. (1999). *Convergence of Probability Measures*, 2nd edn. (John Wiley & Sons, Inc, New York).

Brassard, G. and Bratley, P. (1988). *Algorithmics, Theory and Practice* (Prentice-Hall).

Bratley, P., Fox, B. L. and Schrage, L. E. (1987). *A Guide to Simulation*, 2nd edn. (Springer-Verlag, New York).

Butler, J., Morrice, D. J. and Mullarkey, P. W. (2001). A Multiple attribute utility theory approach to ranking and Selection. *Management Science* **47**, pp. 800–816.

Chan, N. H., Lee, T. C. M. and Peng, L. A. (2010). On nonparametric Local Inference for Density Estimation. *Computational Statistics and Data Analysis* **54**, 2, pp. 509–515.

Chapman, D. G. (1950). Some Two Sample Tests. *Annals of Mathematical Statistics* **21**, pp. 601–606.

Chen, C. H., Lin J., Yücesan, E. and Chick, S. E. (2000). Simulation Budget Allocation for Further Enhancing the Efficiency of Ordinal Optimization. *Journal of Discrete Event Dynamic Systems* **10**, 3, pp. 251–270.

Chen, E. J. (2003). Derivative Estimation with Finite Differences. *Simulation: Transactions of The Society for Modeling and Simulation International* **79**, 10, pp. 598–609.

Chen, E. J. (2004). Using Ordinal Optimization Approach to Improve Efficiency of Selection Procedures. *Discrete Event Dynamic Systems* **14**, 2, pp. 153–170.

Chen, E. J. (2008). Selecting Designs With the Smallest Variance of Normal Populations. *Journal of Simulation* **2**, 3, pp. 186–194.

Chen, E. J. (2011). A Revisit of Two-Stage Selection Procedures. *European Journal of Operational Research* **210**, 2, pp. 281–286.

Chen, E. J. (2012). A Stopping Rule Using the Quasi-Independent Stopping Sequence. *Journal of Simulation* **6**, 2, pp. 71–80.

Chen, E. J. (2013). Some Insights of Using Common Random Numbers in Selection Procedures. *Discrete Event Dynamic Systems* **23**, 3, pp. 241–259.

Chen, E. J. (2014a). Range Statistics and Equivalence Tests. *Journal of Simulation* **8**, 2, pp. 143–150.

Chen, E. J. (2014b). Selection and Order Statistics from Correlated Normal Random Variables. *Discrete Event Dynamic Systems* **24**, 4, pp. 659–668.

Chen, E. J. and Kelton, W. D. (2003). Determining Simulation Run Length with the Runs Test. *Simulation Modelling Practice and Theory* **11**, (3-4), pp. 237–250.

Chen, E. J. and Kelton, W. D. (2005). Sequential Selection Procedures: Using Sample Means to Improve Efficiency. *European Journal of Operational Research* **166**, 2, pp. 133–153.

Chen, E. J. and Kelton, W. D. (2007). A Procedure for Generating Batch-Means Confidence Intervals for Simulation: Checking Independence and Normality. *Simulation: Transactions of The Society for Modeling and Simulation International* **83**, 10, pp. 683–694.

Chen, E. J. and Kelton, W. D. (2008). Estimating Steady-State Distributions via Simulation-Generated Histograms. *Computers and Operations Research* **35**, 4, pp. 1003–1016.

Chen, E. J. and Kelton, W. D. (2010). Confidence-Interval Estimation Using Quasi-Independent Sequences. *IIE Transactions* **42**, 1, pp. 83–93.

Chen, E. J. and Kelton, W. D. (2014). Density Estimation from Correlated Data. *Journal of Simulation* **8**, 4, pp. 281–292.

Chen, E. J. and Lee, L. H. (2014). A Multi-Objective Selection Procedure of Determining a Pareto Set. *Computers & Operations Research* **36**, pp. 1872–

1879

Chen, E. J. and Li, M. (2010) A New Approach to Estimate the Critical Constant of Selection Procedures. *Advances in Decision Sciences*, Volume 2010, Article ID 948359, 12 pages.

Chen, E. J. and Li, M. (2014) Design of Experiments for Interpolation-Based Metamodels. *Simulation Modelling Practice and Theory* **44**, pp. 14–25.

Cheng, R. C. H. (1997). The Generation of Gamma Variables with Non-integral Shape Parameter. *Appl. Statist.* **26**, pp. 71-75.

Chien, C., Goldsman D. and Melamed, B. (1997). Large-Sample Results for Batch Means. *Management Science* **43**, 9, pp. 288–1295.

DasGupta, A. (2011). *Probability for Statistics and Machine Learning* (Springer, New York).

David, H. A. and Nagaraja, H. N. (2003). *Order statistics*, 3d edn. (Wiley, New York).

Devroye, L. J. (1995). *Probability and Statistics for Engineering and the Sciences*, 4th edn. (Brooks/Cole, Monterey, California).

Dudewicz, E. J. and Ahmed, S. U. (1998). New Exact and Asymptotically Optimal Solution to the Behrens-Fisher Problem, With Tables. *American Journal of Mathematical and Management Sciences* **18**, pp. 359–426.

Dudewicz, E. J. and Dalal, S. R. (1975). Allocation of Observations in Ranking and Selection with Unequal Variances. *Sankhya* **B37**, pp. 28–78.

Dudewicz, E. J., Ma, Y., Mai, E. and Su, H. (2007). Exact Solutions to the Behrens-Fisher Problem: Asymptotically Optimal and Finite Sample Efficient Choice Among. *Journal of Statistical Planning and Inference* **137**, pp. 1584–1605.

Dudewicz, E. J., Taneja, V. S. (1978). Multivariate Ranking and Selection Without Reduction to a Univariate Problem, in *Proceedings of the 1978 Winter Simulation Conference*, pp. 207–210.

Duin, R. P. W. (1976). On the Choice of Smoothing Parameters for Parzen Estimators of Probability Density Functions. *IEEE Transactions on Computers* **C-25**, 11, pp. 1175–1179.

Dunnett, C. W and Sobel, M. (1955). Approximations to the Probability Integral and Certain Percentage Points of a Multivariate Analogue of Student's *t*-distribution. *Biometrika* **42**, pp. 258–260.

Edwards, D. G. and Hsu, J. C. (1983). Multiple comparisons with the best treatment. *Journal of the American Statistical Association* **78**, pp. 965-971.

Eickhoff, M., McNickle, D. and Pawlikowski, K. (2006). Analysis of the Time Evolution of Quantiles in Simulation. *International Journal on Simulation: Systems, Science & Technology* **7**, 6, pp. 44–55.

Fishman, G. S. (1978). Grouping observations in digital simulation. *Management Science* **24**, 5, pp. 510-521.

Fishman, G. S. (2001). *Discrete-Event Simulation: Modeling Programming and Analysis.* (Springer-Verlag, New York).

Fujimoto, R. M. (2000). *Parallel and Distributed Simulation Systems.* (Wiley Interscience).

Golyandina, N., Pepelyshev, A. and Steland, A. (2012). New Approaches to Nonparametric Density Estimation and Selection of Smoothing Parameters. *Computational Statistics and Data Analysis* **56**, pp. 2206–2218.

Gupta, S. S., Nagel, K. and Panchapakesan, S. (1973). On the Order Statistics from Equally Correlated Normal Random Variables. *Biometrika* **60**, 2, pp. 403–413.

Hastings, C., Jr. (1955). *Approximations for Digital Computers* (Princeton Univ. Press, Princeton, New Jersey).

Hearne, L. B. and Wegman, E. J. (1994). Fast Multidimensional Density Estimation Based on Random-Width Bins. *Computing Science and Statistics* **26**, pp. 150–155.

Heidelberger, P. and Lewis, P. A. W. (1984). Quantile estimation in dependent sequences. *Operations Research* **32**, pp. 185-209.

Ho, Y. C. (1996). *Soft Optimization for Hard Problems.* (Computerized Lecture Via Private Communication/Distribution).

Ho, Y. C., Sreenivas R. S. and Vakili, P. (1992). Ordinal Optimization of DEDS. *Journal of Discrete Event Dynamic Systems* **2**, pp. 61–68.

Hoad, K., Robinson, S. and Davies, R. (2010). Automating warm-up length estimation. *Journal of the Operational Research Society* **61**, 6, pp. 1389-1403.

Hoad, K., Robinson, S. and Davies, R. (2011). AutoSimOA: a framework for automated analysis of simulation output. *Journal of Simultion* **5**, 1, pp. 9-24.

Hogg, R. V., McKean, J. and Craig, A. T. (2012). *Introduction to Mathematical Statics*, 7th edn. (Pearson Education).

Hurley, C. and Modarres, R. (1995). Low-storage quantile estimation. *Computational Statistics* **10**, pp. 311-325.

Iglehart, D. L. (1976). Simulating stable stochastic systems; VI. quantile estimation. *Journal of the Association of Computing Machinery* **23**, pp. 347-60.

John, T. T. and Chen, P. (2006). Lognormal Selection With Applications to Lifetime Data. *IEEE Transactions on Reliability* **55**, 1, pp. 135–148.

Jones, D. R., Schonlau, M. and Welch, W. J. (1998). Efficient Global Optimization of Expensive Black-Box Functions. *Journal of Glogal Optimization* **13**, pp. 455–492.

Kanter, I., Aviad., Y., Reidler, I., Cohen, E. and Rosenbluh, M. (2010). An optical ultrafast random bit generator. *Nature Photonics* **4**, 1, pp. 58–61.

Knuth, D. E. (1998). *The Art of Computer Programming, Volume 2: Seminumerical Algorithms*, 3rd edn. (Addison-Wesley, Reading, Mass).

Kroese, D. P., Taimre, T. and Botev, Z. I. (2011). *Handbook of Monte Carlo Methods*, (John Wiley & Sons, New York).

L'Ecuyer, P. (1990). Random numbers for simulation. *Communications of the ACM* **33**, 10, pp. 85–97.

L'Ecuyer, P. (1996). Combined multiple recursive random number generators. *Operations Research* **44**, 5, pp. 816–822.

L'Ecuyer, P. (1999). Good parameters and implementations for combined multiple recursive random number generators. *Operations Research* **47**, 1, pp. 159–164.

L'Ecuyer, P., Simard, R., Chen, E. J., Kelton, W. D. (2002). An object-oriented random-number package with many long streams and substreams. *Operations Research* **50**, 6, pp. 1073–1074.

Lada, E. K., Steiger, N. M. and Wilson, J. R. (2008). SBatch: A spaced batch means procedure for steady-state simulation analysis. *Journal of Simulation* **2**, 3, pp. 170-185.

Law, A. M. (2014). *Simulation Modeling and Analysis*, 5th edn. (McGraw-Hill, New York).

Law, A. M. and Carson, J. S. (1979). A Sequential Procedure For Determining the Length of a Steady-State Simulation. *Operations Research* **27**, pp. 1011-1025.

Lee, L. H., Chew, E. P., Teng, S. and Goldsman, D. (2004). Optimal Computing Budget Allocation for Multi-Objective Simulation Models, in *Proceedings of the 2004 Winter Simulation Conference*, pp. 586–594.

Lee, L. H., Chew, E. P. and Teng, S. (2006). Integration of Statistical Selection with Search Mechanism for Solving Multi-Objective Simulation-Optimization Problems, in *Proceedings of the 2006 Winter Simulation Conference*, pp. 294–303.

Liu, W., Ah-Kine, P., Bretz, F. and Hayter, A. J. (2012). Exact Simultaneous Confidence Intervals for a Finite Set of Contrasts of Three, Four or Five Generally Correlated Normal Means. *Computational Statistics and Data Analysis* **57**, pp. 141–148.

Lophaven, S. N., Nielsen, H. B. and Sondergaard, J. (2002) A Matlab Kriging Toolbox, Version 2.5, IMM Technical University of Denmark, Lyngby.

Luo, Y-C, Chen, C. H., Yücesan, E. and Lee, I. (2000). Distributed Web-Based Simulation Optimization, in *Proceedings of the 2000Winter Simulation Conference*, pp. 1785-1793.

Mahamunulu, D. M. (1967). Some Fixed-Sample Ranking and Selection Problems. *Ann Math Stat* **38**, pp. 1079-1091.

Marsaglia, G. and Bray, T. A. (1964). A Convenient Method for Generating Normal Varialbes. *SIAM Review* **6**, 3, pp. 260–264.

Matsumoto, M. and Nishimura, T. (1998). Mersenne twister: A 623-dimensionally equidistributed uniform pseudo-random number generator. *ACM Transactions on Modeling and Computer Simulation* **8**, 1, pp. 3–30.

Meisner, D., Wu, J. and Wenisch, T. F. (2012). BigHouse: A simulation infrastructure for data center systems. *International Symposium on Performance Analysis of Systems and Software (ISPASS)*.

Morrice, D. J. and Butler, J. C. (2006). Ranking and selection with multiple "targets", in *Proceedings of the 2006 Winter Simulation Conference*, pp. 222–230.

Nakayama, M. K. (1997). Multiple-comparison procedures for steady-state simulations. *Annals of Statistics* **25**, pp. 2433-2450.

Nazzal, D., Mollaghasemi, M., Hedlund, H. and Bozorgi, A. (2012). Using genetic algorithms and an indifference-zone ranking and selection procedure under common random numbers for simulation optimisation. *Journal of Simulation* **6**, 1, pp. 56-66.

Pawlikowski, K. (1990). Steady-State Simulation of Queuing Processes: A Survey

of Problems and Solutions. *ACM Computing Surveys* **22**, 2, pp. 123-170.
Prokof'yev, V. N. and Shishkin, A. D. (1974). Successive Classification of Normal Sets with Unknown Variances. *Radio Engng. Electron. Phys.* **19**, 2, pp. 141–143.
Raatikainen, K. E. E. (1990). Sequential procedure for simultaneous estimation of several percentiles. *Transactions of the Society for Computer Simulation* **7**, 1, pp. 21-44.
Rinott, Y. (1978). On two-stage selection procedures and related probability inequalities. *Communications in Statistics* **A7**, 8, pp. 799-811.
Rosenblatt, M. (1971). Curve Estimates. *Annals of Mathematical Statistics* **42**, 6, pp. 1815–1842.
Scheffé, H. (1970). Practical solutions of the Behrens-Fisher Problem. *Journal of American Statistician Association* **65**, pp. 1501–1508.
Schmeiser, B. W. (1982). Batch-size effects in the analysis of simulation output. *Operations Research* **30**, pp. 556-568.
Scott, D. W. and Factor, L. E. (1981). Monte Carlo Study of Three Data-Based Nonparametric Probability Density Estimators. *Journal of the American Statistical Association* **76**, 373, pp. 9–15.
Scott, D. W. and Sain, S. R. (2004). Multi-dimensional Density Estimation in *Handbook of Statistics.* Vol 23: Data Mining and Computational Statistics. eds: CR Rao and EJ Wegman. (Elsevier, Amsterdam).
Scott, D. W., Tapia, R. A. and Thompson, J. R. (1977). Kernel Density Estimation Revisited. *Journal of Nonlinear Analysis, Theory, Methods and Applications* **1**, 4, pp. 339–372.
Seila, A. F. (1982). A Batching approach to quantile estimation in regenerative simulations. *Management Science* 28, 5, pp. 573-81.
Sen, P. K. (1972). On the Bahadur Representation of Sample Quantiles for Sequences of ϕ-mixing Random Variables. *Journal of Multivariate Analysis* **2**, 1, pp. 77–95.
Sheather, S. J. (2004). Density Estimation. *Statistical Science* **19**, 4, 588–597.
Silverman, B. W. (1986). *Density Estimation for Statistics and Data Analysis.* (Chapman and Hall, New York).
Singham D. I. (2014). Selecting Stopping Rules for Confidence Interval Procedure. *ACM Transactions on Modeling and Computer Simulation* **24**, 3, Article No.: 18.
Tafazzoli, A., Steiger, N. M. and Wilson, J. R. (2011). N-Skart: A Nonsequential Skewness-and Autoregression-Adjusted Batch-Means Procedure for Simulation Analysis. *Automatic Control, IEEE Transactions* **45**, 2, pp. 254–264.
Tippett, L. H. C. (1925). On the Extreme Individuals and the Range of Samples Taken from a Normal Population. *Biometrika* **17**, pp. 264–387.
Tong, Y. L. (1980). *Probability Inequalities in Multivarate Distributions.* (Academic Press, New York).
Turner, A. J., Balestrini-Robinson, S. and Mavris, D. (2013). Heuristics for the Regression of Stochastic Simulations *Journal of Simulation* **7**, pp. 229–239.
van Beers, W. C. M. and Kleijnen, J. P. C. (2008). Customized sequential designs for random simulation experiments: Kriging metamodeling and bootstrap-

ping. *European Journal of Operational Research* **186**, pp. 1099–1113.

von Neumann, J. (1941). Distribution of the Ratio of the Mean Square Successive Difference to the Variance. *Annals of Mathematical Statistics* **12**, 4, pp. 367–395.

Welch, B. L. (1938). The Significance of the Difference Between Two Means When the Population Variances are Unequal. *Biometrika* **25**, pp. 350–362.

Wilcox, R. R. (1984). A Table for Rinott's Selection Procedure. *Journal of Quality Technology* **16**, pp. 97–100.

Wilkrmaratna, R. S. (2008). The additive congruential random number generator - A special case of a multiple recursive generator. *Journal of Computational and Applied Mathematics* **216**, 2, pp. 371–387.

Index

Printed in the United States
By Bookmasters